Novel Insights into Orbital Angular Momentum Beams

Novel Insights into Orbital Angular Momentum Beams: From Fundamentals, Devices to Applications

Special Issue Editors

Yang Yue
Hao Huang
Yongxiong Ren
Zhongqi Pan
Alan E. Willner

MDPI • Basel • Beijing • Wuhan • Barcelona • Belgrade

Special Issue Editors
Yang Yue
Nankai University,
China

Hao Huang
Lumentum Operations LLC,
USA

Yongxiong Ren
Gyrfalcon Technologies,
USA

Zhongqi Pan
University of Louisiana at Lafayette,
USA

Alan E. Willner
University of Southern California,
USA

Editorial Office
MDPI
St. Alban-Anlage 66
4052 Basel, Switzerland

This is a reprint of articles from the Special Issue published online in the open access journal *Applied Sciences* (ISSN 2076-3417) from 2018 to 2019 (available at: https://www.mdpi.com/journal/applsci/special_issues/Orbital_Angular_Momentum_Beams)

For citation purposes, cite each article independently as indicated on the article page online and as indicated below:

LastName, A.A.; LastName, B.B.; LastName, C.C. Article Title. *Journal Name* **Year**, *Article Number*, Page Range.

ISBN 978-3-03921-223-1 (Pbk)
ISBN 978-3-03921-224-8 (PDF)

Contents

About the Special Issue Editors

Yang Yue received his B.S. and M.S. degrees in electrical engineering and optics from Nankai University, Tianjin, China, in 2004 and 2007, respectively. He received his Ph.D. degree in electrical engineering from the University of Southern California, Los Angeles, CA, USA, in 2012. He is currently a Professor with the Institute of Modern Optics, Nankai University, Tianjin, China. Dr. Yue's current research interests include intelligent photonics, optical communications and networking, optical interconnect, detection, imaging and display technology. He has published over 130 peer-reviewed journal papers and conference proceedings, 1 edited book, 1 book chapter, >10 invited papers, >30 issued or pending patents, and >60 invited presentations.

Hao Huang received his B.S. degree from Jilin University, Changchun, China, in 2006, and M.S. degree from Beijing University of Posts and Telecommunications, Beijing, China, in 2009. He received his Ph.D. degree in electrical engineering from University of Southern California, Los Angeles, California, USA, in 2014. He is currently working at Lumentum Operations LLC as a hardware engineer. His research area includes optical communication system and components, optical sensing systems, and digital signal processing. He has coauthored more than 100 publications, including peer-reviewed journals and conference proceedings. He is a member of the Optical Society of America (OSA).

Yongxiong Ren received his B.E. degree in Communications Engineering from Beijing University of Posts and Telecommunications (BUPT), Beijing, China, in 2008, and the M.S. degree in Radio Physics from Peking University (PKU), Beijing, China, in 2011. He obtained his Ph.D. degree in electrical engineering from University of Southern California (USC), Los Angeles, California, in 2016. His main research focuses include high-capacity free-space and fiber optical communications, millimeter-wave communications, space division multiplexing, orbital angular momentum multiplexing, atmospheric optics, and atmospheric turbulence compensation. Dr. Ren has authored and coauthored more than 130 research papers with a Google Scholar citation number of >5500. His publications include 57 peer-reviewed journal articles, 76 international conference proceedings, 1 book chapter, and 3 patents.

Zhongqi Pan received his B.S. and M.S. degrees from Tsinghua University, China, and Ph.D. degree from the University of Southern California, Los Angeles, all in Electrical Engineering. He is currently a Professor at the Department of Electrical and Computer Engineering. He also holds the BORSF Endowed Professorship in Electrical Engineering II, and BellSouth/BoRSF Endowed Professorship in Telecommunications. Dr. Pan's research is in the area of photonics, including photonic devices, fiber communications, wavelength-division-multiplexing (WDM) technologies, optical performance monitoring, coherent optical communications, space-division-multiplexing (SDM) technologies, and fiber sensor technologies. He has authored/co-authored 160 publications, including 5 book chapters, 18 invited presentations/papers. He also has 5 U.S. patents and 1 patent in China. Prof. Pan is an OSA and IEEE senior member.

Alan E. Willner received his Ph.D. degree from Columbia University, New York, NY, USA, in 1988. He is currently the Steven and Kathryn Sample Chaired Professor in Engineering with the University of Southern California, Los Angeles, CA, USA. He has more than 1300 publications, including 1 book, 9 edited books, 34 U.S. patents, 39 keynotes/plenaries, 23 book chapters, >375 refereed journal papers, and >250 invited papers/presentations. His research is in optical technologies (e.g., communications, signal processing, networks, and subsystems). He is a member of the U.S. National Academy of Engineering, an International Fellow of the U.K. Royal Academy of Engineering, and a Fellow of the National Academy of Inventors, the American Association for the Advancement of Science, the Optical Society of America, and SPIE.

Editorial

Special Issue on Novel Insights into Orbital Angular Momentum Beams: From Fundamentals, Devices to Applications

Yang Yue [1,*], Hao Huang [2], Yongxiong Ren [2], Zhongqi Pan [3] and Alan E. Willner [2]

1 Institute of Modern Optics, Nankai University, Tianjin 300350, China
2 Department of Electrical Engineering, University of Southern California, Los Angeles, CA 90089, USA
3 Department of Electrical and Computer Engineering, University of Louisiana at Lafayette, Lafayette, LA 70504, USA
* Correspondence: yueyang@nankai.edu.cn

Received: 13 June 2019; Accepted: 24 June 2019; Published: 27 June 2019

1. Introduction

It is well-known now that angular momentum carried by elementary particles can be categorized as spin angular momentum (SAM) and orbital angular momentum (OAM). In the early 1900s, Poynting recognized that a particle, such as a photon, can carry SAM, which has only two possible states, i.e., clockwise and anticlockwise circular polarization states. However, only fairly recently, in 1992, Allen et al. discovered that photons with helical phase fronts can carry OAM, which has infinite orthogonal states [1]. In the past two decades, the OAM-carrying beam, due to its unique features, has gained increasing interest from many different research communities, including physics, chemistry and engineering [2,3]. Its twisted phase front and intensity distribution have enabled a variety of applications, such as micromanipulation [4–6], laser beam machining [7–9], nonlinear matter interactions [10–12], imaging [13–15], sensing [16,17], quantum cryptography, and classical communications [18–23].

2. Special Issue Papers

This special issue aims to explore the novel insights of OAM beams. It focuses on state-of-the-art advances in fundamental theories, devices, and applications as well as future perspectives of OAM beams. The collected papers have well accomplished these goals by contributing leading-edge derivation, analysis, and experiments with significant results. The topics cover OAM generation and reception, multiplexing and de-multiplexing, device and system. The frequencies range from radio frequency (RF) to infrared wave, while the techniques behind extend from integrated photonics, fiber optics, free-space optics, to dielectric. The special issue consists of three review papers, one communication and five research articles.

More specifically, from the physical perspective, Prof. Barnett and his group have a review paper on the helicity of light, and how it can be both produced and used in light-matter interactions [24]. The paper starts from the form of the helicity density and its associated continuity equation in free space, in the presence of local currents and charges, and upon interaction with bulk media, leading to the characterization of both microscopic and macroscopic sources of helicity.

Regarding OAM beam generation technologies, Prof. Liu's group reviews the generation of OAM modes using fiber systems [25]. This review paper first introduces the basic concepts of fiber modes and the generation and detection theories of OAM modes. Next, fiber systems based on different devices are introduced, including long-period fiber grating, mode-selective coupler, microstructured optical fiber, and the photonic lantern. Finally, the key challenges and prospects for fiber OAM mode systems are discussed.

Another review paper, focusing on tunable OAM generation, is contributed from Prof. Wu's group [26]. The authors classify the tunable OAM mode generation methods into three categories, according to the OAM and polarization states. The fiber-based and free-space generation methods are categorized into three types according to the controllable variables, respectively. Last, the pros and cons of each generation method are analyzed and the key challenges for tunable OAM modes are discussed.

Most fiber-based or free-space OAM beam generators are bulky, slow, and cannot withstand high powers. Prof. Litchinitser's group design and experimentally demonstrate an ultra-fast, compact chalcogenide-based all-dielectric metasurface beam converter, which has the ability to transform a Hermite–Gaussian (HG) beam into an OAM beam at near infrared wavelength [27]. The topological charge carried by the output OAM beam can be switched between positive and negative values, and the device provides high transmission efficiency.

For the reception of OAM radio waves, Dr. Klemes contributes a communication article using pseudo-Doppler interpolation techniques [28]. The method can be used to receive OAM waves in the far field of an antenna transmitting multiple OAM modes, each carrying a separate data stream at the same RF. The frequency domain method provides a higher signal-to-noise ratio (SNR) than using spatial-domain OAM reception techniques. Moreover, no more than two receiving antennas are necessary to separate any number of OAM modes in principle.

In OAM communications systems, different OAM beams can carry multiple data channels, boosting the spectral efficiency and capacity significantly. Consequently, the simultaneous processing of OAM beams is necessary, and OAM multiplexing/de-multiplexing devices are key enablers of such systems. Prof. Li et al. contribute an article on mode-selective photonic lanterns for OAM mode division multiplexing [29]. The authors design a three-mode OAM mode-selective photonic lantern by optimizing the taper length with small mode crosstalk, which employs only a single mode fiber port to selectively generate each OAM mode.

In a more integrated manner, Prof. Romanato and his group explore holographic Silicon metasurfaces for OAM de-multiplexing based on OAM-mode projection [30]. The device uses Pancharatnam-Berry optical elements (PBOEs) and can de-multiplex beams with different polarization and OAM states at the wavelength of 1310 nm. The geometric-phase control is achieved by inducing a spatially-dependent form-birefringence on a silicon substrate, patterned with properly-oriented subwavelength gratings.

There are two experimental demonstrations for free-space OAM communications systems, one in the 1550 nm optical regime, and the other on 28 GHz RF frequency band. Dr. Qu and Prof. Djordjevic investigate turbulence mitigation methods in free-space optical OAM communications system based on coded modulation [31]. Adaptive optics, channel coding, Huffman coding combined with low-density parity-check (LDPC) coding, and spatial offset are used for turbulence mitigation, achieving a total data transmission capacity of 500 Gbps.

Finally, Dr. Lee and colleagues evaluate the performance of OAM-based wireless communications systems [32]. To overcome the beam divergence of OAM multiplexing, the authors use a combination of multi-input multi-output (MIMO) and OAM technology, achieving a new milestone in point-to-point transmission rates at 100 Gbps for a 10 m transmission distance.

3. Perspectives

It has just been 27 years since the discovery of OAM by Les Allen and his co-workers. Within this fairly short period of time, an extensive research community has been established globally, and OAM theories have been further improved. Especially during the past decade, OAM related devices and applications have experienced significant growth. From this trend, we are expecting the OAM field to continue grow with novel and unique applications to debut one after another. Hopefully, more OAM related technologies can be commercialized in the near future to enable new industry and serve society.

Acknowledgments: The guest editors would like to thank all the authors for submitting their excellent work to this special issue. Furthermore, we would like to thank all the reviewers for their outstanding job in evaluating the manuscripts and providing helpful comments. The guest editors also would like to thank the MDPI team involved in the preparation, editing, and managing of this special issue. Finally, we would like to express our sincere gratitude to Ms. Lucia Li, the contact editor of this special issue, for her kind, efficient, professional guidance and support through the whole process. We would not be able to reach the above collection of high quality papers without this joint effort.

Conflicts of Interest: The authors declare no conflict of interest.

References

1. Allen, L.; Beijersbergen, M.W.; Spreeuw, R.J.C.; Woerdman, J.P. Orbital angular momentum of light and the transformation of Laguerre-Gaussian laser modes. *Phys. Rev. A* **1992**, *45*, 8185–8189. [CrossRef] [PubMed]
2. Yao, A.M.; Padgett, M.J. Orbital angular momentum: Origins, behavior and applications. *Adv. Opt. Photon.* **2011**, *3*, 161–204. [CrossRef]
3. Padgett, M.J. Orbital angular momentum 25 years on [Invited]. *Opt. Express* **2017**, *25*, 11265–11274. [CrossRef] [PubMed]
4. Friese, M.E.J.; Nieminen, T.A.; Heckenberg, N.R.; Rubinsztein-Dunlop, H. Optical alignment and spinning of laser-trapped microscopic particles. *Nature* **1998**, *394*, 348–350. [CrossRef]
5. Dholakia, K.; Cizmar, T. Shaping the future of manipulation. *Nat. Photonics* **2011**, *5*, 335–342. [CrossRef]
6. Padgett, M.; Bowman, R. Tweezers with a twist. *Nat. Photonics* **2011**, *5*, 343–348. [CrossRef]
7. Friese, M.; Rubinsztein-Dunlop, H.; Gold, J.; Hagberg, P.; Hanstorp, D. Optically driven micromachine elements. *Appl. Phys. Lett.* **2001**, *78*, 547–549. [CrossRef]
8. Knoner, G.; Parkin, S.; Nieminen, T.A.; Loke, V.L.Y.; Heckenberg, N.R.; Rubinsztein-Dunlop, H. Integrated optomechanical microelements. *Opt. Express* **2007**, *15*, 5521–5530. [CrossRef]
9. Ladavac, K.; Grier, D. Micro-optomechanical pumps assembled and driven by holographic optical vortex arrays. *Opt. Express* **2004**, *12*, 1144–1149. [CrossRef]
10. Firth, W.; Skryabin, D. Optical solitons carrying orbital angular momentum. *Phys. Rev. Lett.* **1997**, *79*, 2450–2453. [CrossRef]
11. Litchinitser, N.M. Applied physics. Structured light meets structured matter. *Science* **2012**, *337*, 1054–1055. [CrossRef] [PubMed]
12. Gauthier, D.; Ribic, P.R.; Adhikary, G.; Camper, A.; Chappuis, C.; Cucini, R.; DiMauro, L.F.; Dovillaire, G.; Frassetto, F.; Géneaux, R.; et al. Tunable orbital angular momentum in high-harmonic generation. *Nat. Commun.* **2017**, *8*, 14971. [CrossRef]
13. Hell, S.W.; Wichmann, J. Breaking the diffraction resolution limit by stimulated emission: Stimulated-emission-depletion fluorescence microscopy. *Opt. Lett.* **1994**, *19*, 780–782. [CrossRef]
14. Swartzlander, G.A.; Ford, E.L.; Abdul-Malik, R.S.; Close, L.M.; Peters, M.A.; Palacios, D.M.; Wilson, D.W. Astronomical demonstration of an optical vortex coronagraph. *Opt. Express* **2008**, *16*, 10200–10207. [CrossRef] [PubMed]
15. Jesacher, A.; Ritsch-Marte, M.; Piestun, R. Three-dimensional information from two-dimensional scans: A scanning microscope with postacquisition refocusing capability. *Optica* **2015**, *2*, 210–213. [CrossRef]
16. Cvijetic, N.; Milione, G.; Ip, E.; Wang, T. Detecting lateral motion using light's orbital angular momentum. *Sci. Rep.* **2015**, *5*, 15422. [CrossRef] [PubMed]
17. Xie, G.; Song, H.; Zhao, Z.; Milione, G.; Ren, Y.; Liu, C.; Zhang, R.; Bao, C.; Li, L.; Wang, Z.; et al. Using a complex optical orbital-angular-momentum spectrum to measure object parameters. *Opt. Lett.* **2017**, *42*, 4482–4485. [CrossRef] [PubMed]
18. Wang, J.; Yang, J.Y.; Fazal, I.M.; Ahmed, N.; Yan, Y.; Huang, H.; Ren, Y.X.; Yue, Y.; Dolinar, S.; Tur, M.; et al. Terabit free-space data transmission employing orbital angular momentum multiplexing. *Nat. Photonics* **2012**, *6*, 488–496. [CrossRef]
19. Bozinovic, N.; Yue, Y.; Ren, Y.; Tur, M.; Kristensen, P.; Huang, H.; Willner, A.E.; Ramachandran, S. Terabit-scale orbital angular momentum mode division multiplexing in fibers. *Science* **2013**, *340*, 1545–1548. [CrossRef]
20. Krenn, M.; Handsteiner, J.; Fink, M.; Fickler, R.; Ursin, R.; Malik, M.; Zeilinger, A. Twisted light transmission over 143 km. *Proc. Natl. Acad. Sci. USA* **2013**, *113*, 13648–13653. [CrossRef]

21. Yan, Y.; Xie, G.; Lavery, M.P.J.; Huang, H.; Ahmed, N.; Bao, C.; Ren, Y.; Cao, Y.; Li, L.; Zhao, Z.; et al. High-capacity millimetre-wave communications with orbital angular momentum multiplexing. *Nat. Commun.* **2014**, *5*, 4876. [CrossRef] [PubMed]
22. Vallone, G.; D'Ambrosio, V.; Sponselli, A.; Slussarenko, S.; Marrucci, L.; Sciarrino, F.; Villoresi, P. Free-space quantum key distribution by rotation-invariant twisted photons. *Phys. Rev. Lett.* **2014**, *113*, 060503. [CrossRef] [PubMed]
23. Sit, A.; Bouchard, F.; Fickler, R.; Gagnon-Bischoff, J.; Larocque, H.; Heshami, K.; Elser, D.; Peuntinger, C.; Günthner, K.; Heim, B.; et al. High-dimensional intracity quantum cryptography with structured photons. *Optica* **2017**, *4*, 1006–1010. [CrossRef]
24. Crimin, F.; Mackinnon, N.; Götte, J.B.; Barnett, S.M. Optical helicity and chirality: Conservation and sources. *Appl. Sci.* **2019**, *9*, 828. [CrossRef]
25. Zhang, H.; Mao, B.; Han, Y.; Wang, Z.; Yue, Y.; Liu, Y. Generation of orbital angular momentum modes using fiber systems. *Appl. Sci.* **2019**, *9*, 1033. [CrossRef]
26. Feng, L.; Li, Y.; Wu, S.; Li, W.; Qiu, J.; Guo, H.; Hong, X.; Zuo, Y.; Wu, J. A review of tunable orbital angular momentum modes in fiber: Principle and generation. *Appl. Sci.* **2019**, *9*, 2408. [CrossRef]
27. Xu, Y.; Sun, J.; Frantz, J.; Shalaev, M.I.; Walasik, W.; Pandey, A.; Myers, J.D.; Bekele, R.Y.; Tsukernik, A.; Sanghera, J.S.; et al. Nonlinear metasurface for structured light with tunable orbital angular momentum. *Appl. Sci.* **2019**, *9*, 958. [CrossRef]
28. Klemes, M. Reception of OAM radio waves using pseudo-doppler interpolation techniques: A frequency-domain approach. *Appl. Sci.* **2019**, *9*, 1082. [CrossRef]
29. Li, Y.; Li, Y.; Feng, L.; Yang, C.; Li, W.; Qiu, J.; Hong, X.; Zuo, Y.; Guo, H.; Tong, W.; et al. Mode-selective photonic lanterns for orbital angular momentum mode division multiplexing. *Appl. Sci.* **2019**, *9*, 2233. [CrossRef]
30. Ruffato, G.; Massari, M.; Capaldo, P.; Romanato, F. Holographic silicon metasurfaces for total angular momentum demultiplexing applications in telecom. *Appl. Sci.* **2019**, *9*, 2387. [CrossRef]
31. Qu, Z.; Djordjevic, I.B. Orbital angular momentum multiplexed free-space optical communication systems based on coded modulation. *Appl. Sci.* **2019**, *9*, 2179. [CrossRef]
32. Lee, D.; Sasaki, H.; Fukumoto, H.; Yagi, Y.; Shimizu, T. An evaluation of orbital angular momentum multiplexing technology. *Appl. Sci.* **2019**, *9*, 1729. [CrossRef]

Review

Optical Helicity and Chirality: Conservation and Sources

Frances Crimin *, Neel Mackinnon, Jörg B. Götte and Stephen M. Barnett

School of Physics and Astronomy, University of Glasgow, Glasgow G12 8QQ, UK;
n.mackinnon.1@research.gla.ac.uk (N.M.); Joerg.Goette@glasgow.ac.uk (J.B.G.);
Stephen.Barnett@glasgow.ac.uk (S.M.B.)
* Correspondence: frances.crimin@glasgow.ac.uk

Received: 14 February 2019; Accepted: 21 February 2019; Published: 26 February 2019

Abstract: We consider the helicity and chirality of the free electromagnetic field, and advocate the former as a means of characterising the interaction of chiral light with matter. This is in view of the intuitive quantum form of the helicity density operator, and of the dual symmetry transformation generated by its conservation. We go on to review the form of the helicity density and its associated continuity equation in free space, in the presence of local currents and charges, and upon interaction with bulk media, leading to characterisation of both microscopic and macroscopic sources of helicity.

Keywords: helicity; chirality; orbital angular momentum; dual symmetry; light–matter interactions; bi-isotropic media

1. Introduction

The study of the handedness, or chirality, of matter has its roots in the work of Arago [1], Biot [2] and Pasteur [3,4] in the early 19th Century, with the discovery that the polarisation of light rotates upon propagation through certain crystals and molecular solutions. It was Lord Kelvin who introduced the word chiral to describe such matter, which is non-superimposable upon its mirror image [5], with a chiral object and its mirror image being called enantiomers. In particular, it was realised in these early experiments that if the rotation angle of the polarisation vector through a solution of chiral molecules is given by θ, then the rotation of the light in the enantiomeric form of the solution is through the angle $-\theta$. We need not look far to observe that chirality is in fact ubiquitous in nature: our left and right hands and feet are distinct from each other, with the word "chiral" itself derived from the Greek word for hand, $\chi\epsilon\iota\rho$ [6]. The weak force, being parity violating, is a striking example of the role of handedness in nature [7], as is the remarkable selectivity evident in biological homochirality [8]: the complex molecules DNA, RNA, as well as the proteins and sugars comprising all living organisms, are indeed chiral. A simple example of a chiral molecule, alongside an achiral counterpart, is shown in Figure 1.

The response of such chiral matter to the polarisation of light is called natural optical activity [9], and it is with this topic which we are primarily concerned. Of course, light itself can have a chiral structure: left- and right-circularly polarised fields trace out helices with opposite handedness, and accordingly act as a chiral influence on matter which itself exists in two enantiomeric forms. This is manifest in many different effects [6], the most well-known being optical rotation, due to left- and redright-circularly polarised light having a different refractive index in the chiral medium [2], and circular dichroism, arising from the different absorption coefficients for the two polarisations [10]. We therefore look for a way of characterising such polarised fields before and after interaction with an object which will allow us to infer any chiral influence of that object. Light carries an intrinsic angular momentum [11], or spin, which differs in sign for left- and right-circular polarisation, promoting the spin as a natural candidate for our purposes [6]. Being a pseudovector, however, means that the spin has even parity, whereas a little thought on the problem reveals that an odd-parity observable is required to distinguish the interaction of light with two enantiomers of a chiral object, which are parity odd. The answer lies in the projection of the spin in the direction of propagation of the light beam: the optical helicity [12–19]. We use the well-known method of constructing continuity equations for conserved quantities in electrodynamics [17] to show that the helicity can be used to characterise the interaction of light with different types, and indeed scales, of matter. In Section 2, we review the definition of the electromagnetic helicity in a vacuum, before looking at microscopic sources of helicity in the form of mixed radiating electric-magnetic dipoles in Section 3. We extend the method in Section 4 to examine the conditions under which helicity is conserved in lossless bulk media, before summarising the results in Section 5.

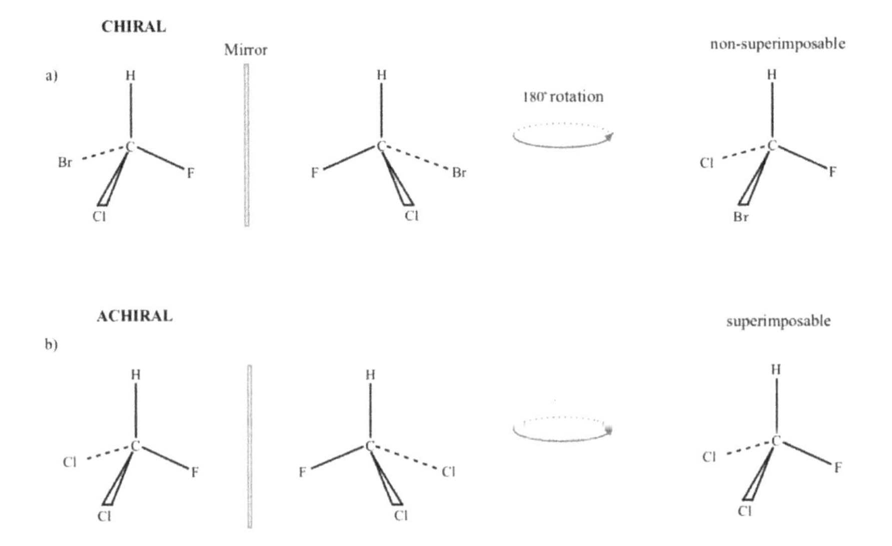

Figure 1. (**a**) The molecule bromochlorofluoromethane is chiral, as the molecule and its mirror image cannot be superimposed, even after rotation. (**b**) Dichlorofluoromethane, on the other hand, is achiral, as the molecule can be superimposed upon its mirror image after rotation.

2. Review and Motivation

In many areas of physics, it is useful to characterise the "twist" of a vector field by a pseudoscalar quantity of the form $\mathbf{F} \cdot (\nabla \times \mathbf{F})$. In fluid mechanics, for example, the quantity $\nabla \times \mathbf{v}$ is called the "vorticity" [20,21], where $\mathbf{v}$ is the fluid flow velocity, and the further quantity $\mathbf{v} \cdot (\nabla \times \mathbf{v})$ describes the knottedness of the vortex lines [22].

In plasma physics, the conserved quantity $\mathbf{A} \cdot (\nabla \times \mathbf{A})$ has been used since the late 1950s to characterise the topology of magnetic field lines [23,24], where $\mathbf{A}$ is the magnetic vector potential. More recently, this quantity has been applied to the study of optical fields. It can be generalised in order to include both electric and magnetic contributions by including an additional vector potential $\mathbf{C}$, defined such that $\nabla \times \mathbf{C} = -\mathbf{D}^{\mathrm{T}}$, where $\mathbf{D}^{\mathrm{T}}$ is the transverse part of the displacement field [25]. For the free field, $\mathbf{D}^{\mathrm{T}} = \epsilon_0 \mathbf{E}$, and the helicity density is often expressed in terms of $\mathbf{E}$ rather than $\mathbf{D}$. In the presence of a medium, however, it is preferable to use the form given here. This leads to the symmetrical definition:

$$h = \frac{1}{2}\left[\sqrt{\frac{\epsilon_0}{\mu_0}}\mathbf{A} \cdot (\nabla \times \mathbf{A}) + \sqrt{\frac{\mu_0}{\epsilon_0}}\mathbf{C} \cdot (\nabla \times \mathbf{C})\right]. \tag{1}$$

This is the quantity we refer to as the "helicity" (or, more accurately, "helicity density") throughout this article. The helicity is a Lorentz pseudoscalar and is a conserved quantity of the free electromagnetic field. It is closely connected to the spin angular momentum of light, and has attracted attention as a way of describing the interactions of light with chiral matter [16–18,26].

2.1. Integrated Helicity and Local Densities

Here, we will say a few words about the purpose of extending the magnetic definition of helicity to include the extra gauge potential **C**, leading to the second term in (1). Indeed, the quantity $\int \mathbf{A} \cdot \mathbf{B}\, d^3\mathbf{r}$, where **B** is the magnetic induction, is sometimes used as a measure of the total electromagnetic helicity [27,28], as, when time averaged, it is equivalent to the volume integral of (1). There is a sense in which the helicity is only meaningful when integrated over all space: the appearance of the gauge potentials in the definition implies that the helicity density at a point is explicitly gauge-dependent. However, the total helicity is physically meaningful, as the volume integral serves to pick out only the transverse parts of **A** and **C**, which are gauge-invariant, meaning that the integrated helicity is in fact a gauge-independent quantity [14].

Such gauge-related ambiguity in the definition of the local densities might suggest that the physically meaningful content of both definitions is the same. On the other hand, the manner in which the total integral becomes gauge-independent might suggest that a local helicity density could be unambiguously defined by explicitly using only the transverse parts of the potentials in (1). This is the approach adopted throughout this article, though it should be cautioned that such a helicity density still retains an element of a non-local character, as even the gauge-invariant parts of the potentials at a point are not only determined by the fields and their derivatives at that point; the values of the fields at other points are involved as well [17].

Having said this, there are more than simply aesthetic reasons that the symmetrical definition of helicity might be preferred over the asymmetrical. For one thing, we will see that the symmetrical definition obeys an exact local continuity equation and is therefore locally conserved, at least in the absence of matter [6,17]. Furthermore, it retains its form under both Lorentz and duality transformations [14], the latter being transformations that rotate electric into magnetic fields, and vice versa, encapsulating the symmetry between these fields in the absence of charges. An analogous continuity equation cannot be drawn up for the asymmetrical definition [17], and only its integral over all space is invariant under duality and Lorentz transformations. This feature is the origin of subtle complications in the use of $\mathbf{A} \cdot \mathbf{B}$ in the study of plasmas [24], but such issues do not arise for the locally-conserved form. Finally, as will be shown in Section 2.3, the quantum helicity operator derived from (1) has a particularly intuitive form, in further support of this symmetric definition.

2.2. Helicity and Chirality

The helicity is closely related to the "chirality", another conserved quantity of the free electromagnetic field, introduced by Lipkin [29] as one of a class of conserved quantities called "zilches". It is defined as:

$$\chi = \frac{\epsilon_0}{2}\left[\mathbf{E} \cdot (\nabla \times \mathbf{E}) + c^2 \mathbf{B} \cdot (\nabla \times \mathbf{B})\right]. \tag{2}$$

This quantity has also been studied as a means of describing the interaction of light with chiral matter. Tang and Cohen [30] demonstrated the physical significance of the chirality, showing that the differential excitation rate between the two enantiomers of a chiral molecule in a monochromatic optical field is proportional to the chirality of the field in which they are immersed. Since that work, the chirality density has been applied to the analysis of a number of scenarios, including that of fields near metamaterial surfaces, where configurations have been suggested in which the ratio of chirality density to energy density is greater than in circularly polarised light. This "superchirality" has been proposed as a means of enhancing the enantioselectivity of some chiroptical techniques [31,32], although the importance of such "superchirality" in explaining the reported enhancements has been challenged [33]. Superchirality in the works of Tang and Cohen [31] is associated with regions of destructive interference and hence a reduction in the energy density; however, it is also possible to generate bright regions of superchirality [34].

In monochromatic fields, the helicity and chirality densities are proportional to one another [14]. However, they have different frequency dependencies, and so, in a general time-dependent field, no strict proportionality holds. A simple and striking illustration of the non-equivalence of helicity and chirality is provided by considering the superposition of two circular plane waves of opposite handedness and different frequencies; in this case, the two quantities have opposite signs. For right- and left-circular polarisations, the helicity is proportional to $\pm 1/\omega$ and the chirality to $\pm\omega$. This might immediately suggest that the two measures give opposite signs: if the right-handed wave has a higher frequency than the left-handed, then the magnitude of the chirality of the right-handed wave will be greater than that of the left, but the magnitude of the helicity will be greater for the left than the right. Of course, the helicity and chirality are both quadratic in the fields, so the above represents little more than a plausibility argument; explicit calculation, however, reveals that they do indeed have opposite signs. If we take our two plane waves to be travelling in the positive z direction, then choosing vector potentials $\mathbf{A} = -\int \mathbf{E}\,dt$ and $\mathbf{C} = -\frac{1}{c^2}\int \mathbf{B}\,dt$ and using the real parts of the fields and potentials in (2) and (1) gives:

$$\chi = \frac{2\epsilon_0 E_0^2}{c}(\omega_1 - \omega_2)\cos^2\left(\frac{\omega_1 - \omega_2}{2}\eta\right) \tag{3}$$

and:

$$h = -\frac{2\epsilon_0 E_0^2}{\omega_1\omega_2}(\omega_1 - \omega_2)\cos^2\left(\frac{\omega_1 - \omega_2}{2}\eta\right), \tag{4}$$

where ω_1 is the frequency of the right-handed wave, ω_2 that of the left-handed, E_0 the peak electric field strength of each wave, and $\eta = t - z/c$. We see that the helicity and chirality densities are here both proportional to the energy density at all times, but with opposite signs and different dependencies on the two frequencies.

Given the sign difference, it seems reasonable to ask which of the helicity and chirality corresponds to the intuitive "sense" of the rotation in the example field given here. This field does have a clear intuitive sense of rotation: if there is no frequency difference between the two circular plane waves, the resulting superposition is simply a linearly polarised plane wave, but for small frequency differences, the result is approximately a linearly polarised wave with a frequency equal to the average frequency and a plane of polarisation that slowly rotates at $(\omega_1 - \omega_2)/2$. This rotation of the polarisation plane is directly analogous to the amplitude modulation in the "beats" observed in the addition of two linearly polarised waves of different frequency. Our simple example may also be of some practical interest, as these fields form the basis of an "optical centrifuge", a procedure that can be used to excite very high rotational states in molecules [35]. This is accomplished by introducing a linear chirp into each of the two waves, one chirped up and the other down, so that the speed of rotation increases with time, driving the molecule up a ladder of rotational transitions.

The rotation of the polarisation plane is in the same sense as that of the higher-frequency circularly polarised wave. Therefore, insofar as the sign of these quantities should be a guide to the sense of "rotation" in the field, the chirality density might appear to assign the correct sign (the chirality of the combined field has the same sign as the chirality of the higher frequency wave), while the helicity has the opposite (the helicity of the combined field has the opposite sign to that of the higher frequency wave). However, the angular momentum content of the combined field is actually dominated by the lower frequency plane wave. This can be shown by a simple photon-counting argument: if the two plane waves have the same field strength (and therefore the same energy), then there are more photons in the lower frequency mode than in the higher, as the photons in the lower frequency mode have less energy each. As each photon in the higher frequency mode carries an angular momentum of $\hbar$ and each in the lower frequency mode an angular momentum of $-\hbar$, it is clear that the sign of total angular momentum is indeed reflected by the helicity, rather than the chirality.

2.3. Quantum Helicity Operator

One reason to prefer the helicity over chirality as a measure of the degree of handedness of an optical field is its intuitive quantum-mechanical form. The operators for the electric and magnetic fields and the two vector potentials are given by [17]:

$$\begin{aligned}
\hat{\mathbf{E}} &= \sqrt{\frac{\hbar c k}{2\epsilon_0 V}} \sum_{\mathbf{k},\lambda} i \mathbf{e}_{\mathbf{k},\lambda} \hat{a}_{\mathbf{k},\lambda} e^{i(\mathbf{k}\cdot\mathbf{x}-\omega t)} + h.c, \\
\hat{\mathbf{B}} &= \sqrt{\frac{\hbar}{2\epsilon_0 c k V}} \sum_{\mathbf{k},\lambda} i (\mathbf{k} \times \mathbf{e}_{\mathbf{k},\lambda}) \hat{a}_{\mathbf{k},\lambda} e^{i(\mathbf{k}\cdot\mathbf{x}-\omega t)} + h.c, \\
\hat{\mathbf{A}} &= \sqrt{\frac{\hbar}{2\epsilon_0 c k V}} \sum_{\mathbf{k},\lambda} \mathbf{e}_{\mathbf{k},\lambda} \hat{a}_{\mathbf{k},\lambda} e^{i(\mathbf{k}\cdot\mathbf{x}-\omega t)} + h.c, \\
\hat{\mathbf{C}} &= \sqrt{\frac{\hbar \epsilon_0 c}{2k^3 V}} \sum_{\mathbf{k},\lambda} (\mathbf{k} \times \mathbf{e}_{\mathbf{k},\lambda}) \hat{a}_{\mathbf{k},\lambda} e^{i(\mathbf{k}\cdot\mathbf{x}-\omega t)} + h.c,
\end{aligned} \tag{5}$$

where λ labels two orthogonal polarisation modes, $\mathbf{e}_{\mathbf{k},\lambda}$ and $a_{\mathbf{k},\lambda}$ are respectively the polarisation vector and annihilation operator corresponding to a photon in mode $(\mathbf{k},\lambda)$, V is the quantisation volume and $k \equiv \sqrt{\mathbf{k}^2}$. Using these field and vector potential operators, expanded in the basis of creation and annihilation operators for left- and right-circular polarisation modes, the quantum-mechanical version of the classical definition of the integrated helicity density (1) corresponds to the difference in the total number of left- and right-circularly polarised photons in the field [14,15,17]:

$$\hat{\mathcal{H}} = \int \hat{h}\, d^3 r = \hbar \sum_{\mathbf{k}} \left(\hat{N}_{\mathbf{k},+} - \hat{N}_{\mathbf{k},-} \right). \tag{6}$$

Here, $\hat{N}_{\mathbf{k},+}$ and $\hat{N}_{\mathbf{k},-}$ are the number operators for photons in the $(\mathbf{k},+)$ and $(\mathbf{k},-)$ modes, $\hat{N}_{\mathbf{k},+} \equiv \hat{a}^\dagger_{\mathbf{k},+}\hat{a}_{\mathbf{k},+}$ and $\hat{N}_{\mathbf{k},-} \equiv \hat{a}^\dagger_{\mathbf{k},-}\hat{a}_{\mathbf{k},-}$. This form of the helicity operator allows the electromagnetic definition of helicity to be connected to the concept of helicity in particle physics, where the helicity of a particle is defined as the projection of its spin angular momentum in the direction of propagation [19]. The angular momentum associated classically with circular polarisation is conventionally associated with the photon spin, meaning that each photon in a right- or left-handed mode contributes $\pm\hbar$ of helicity. Parenthetically, it should be noted that the terminology of separating the total angular momentum into spin and orbital "angular momenta" must be approached with caution. There is a sense in which neither the spin, nor the orbital quantum operators correspond to angular momenta, as neither satisfy the commutation relations of an angular momentum [36–39]. However, the quantities are physically distinct, as supported by experimental evidence [39], and are furthermore separately conserved under separate rotational transformations of the electromagnetic fields [36–38,40].

2.4. Helicity and Duality Symmetry

The free-space Maxwell equations treat electric and magnetic fields on equal footing. To say this more precisely: the form of the free space Maxwell equations is invariant under the duality transformation [41]:

$$\begin{aligned}
\mathbf{E} &\longrightarrow \mathbf{E}\cos\theta + c\mathbf{B}\sin\theta, \\
c\mathbf{B} &\longrightarrow c\mathbf{B}\cos\theta - \mathbf{E}\sin\theta,
\end{aligned} \tag{7}$$

for any real (pseudoscalar) angle θ. This property of the free electromagnetic field has been variously referred to as the Heaviside–Larmor symmetry, duplex symmetry [14], dual-symmetry [18] or "electric-magnetic democracy" [42]. The helicity of the electromagnetic field is fundamentally connected with the dual-symmetry of the free-space Maxwell equations: the duality transformation can be obtained by taking the helicity as the generator of an infinitesimal transformation of the fields [14]. To put the matter the other way around, the conservation of helicity in a vacuum can be derived from the dual-symmetry of the free-space Maxwell equations using Noether's theorem [12].

Here, we will examine the approach taken in [15,17], where the quantum-mechanical optical helicity $\hat{h}$ is used to form the transformation operator:

$$\hat{U}(\theta) = \exp\left(-\frac{i}{\hbar}\theta\hat{h}\right), \tag{8}$$

which can be applied to the vector fields $\hat{\mathbf{E}}$ and $\hat{\mathbf{B}}$ in (5) to produce:

$$\begin{aligned} \hat{U}^{\dagger}(\theta)\,\hat{\mathbf{E}}\,\hat{U}(\theta) &= \hat{\mathbf{E}}\cos\theta + c\hat{\mathbf{B}}\sin\theta, \\ \hat{U}^{\dagger}(\theta)\,c\hat{\mathbf{B}}\,\hat{U}(\theta) &= c\hat{\mathbf{B}}\cos\theta - \hat{\mathbf{E}}\sin\theta. \end{aligned} \tag{9}$$

This is in complete analogy with (7), where now, the helicity operator $\hat{h}$ is explicitly shown to generate this transformation. Results for the potentials $\hat{\mathbf{A}}$ and $\hat{\mathbf{C}}$ in (5) follow similarly:

$$\begin{aligned} \hat{U}^{\dagger}(\theta)\,\hat{\mathbf{A}}\,\hat{U}(\theta) &= \hat{\mathbf{A}}\cos\theta + \sqrt{\frac{\mu}{\epsilon}}\hat{\mathbf{C}}\sin\theta, \\ \hat{U}^{\dagger}(\theta)\,\hat{\mathbf{C}}\,\hat{U}(\theta) &= \hat{\mathbf{C}}\cos\theta - \sqrt{\frac{\epsilon}{\mu}}\hat{\mathbf{A}}\sin\theta. \end{aligned} \tag{10}$$

2.3. Continuity Equations in Free Space

As mentioned above, helicity is a conserved quantity in vacuum. This fact can be expressed using a local continuity equation, relating the time derivative of the helicity density at a point to the helicity flux through an infinitesimal volume surrounding that point. Taking the time derivative of the helicity density (1):

$$\begin{aligned} \partial_t h &= \frac{1}{2}\left[\sqrt{\frac{\epsilon_0}{\mu_0}}\left(\dot{\mathbf{A}}\cdot\mathbf{B} + \mathbf{A}\cdot\dot{\mathbf{B}}\right) - \sqrt{\frac{\mu_0}{\epsilon_0}}\left(\dot{\mathbf{C}}\cdot\mathbf{D} + \mathbf{C}\cdot\dot{\mathbf{D}}\right)\right] \\ &= \frac{1}{2}\left[\sqrt{\frac{\epsilon_0}{\mu_0}}\left(-\mathbf{E}\cdot(\nabla\times\mathbf{A}) - \mathbf{A}\cdot(\nabla\times\mathbf{E})\right) - \sqrt{\frac{\mu_0}{\epsilon_0}}\left(\mathbf{H}\cdot(\nabla\times\mathbf{C}) + \mathbf{C}\cdot(\nabla\times\mathbf{H})\right)\right], \end{aligned} \tag{11}$$

where, from the free-space Maxwell equations, we have used $\mathbf{E} = -\dot{\mathbf{A}}$ and $\mathbf{H} = -\dot{\mathbf{C}}$, with the dotted notation indicating the time derivative of the fields. Using the vector identity $\nabla\cdot(\mathbf{E}\times\mathbf{A}) = \mathbf{A}\cdot(\nabla\times\mathbf{E}) - \mathbf{E}\cdot(\nabla\times\mathbf{A})$ and $\nabla\cdot(\mathbf{H}\times\mathbf{C}) = \mathbf{C}\cdot(\nabla\times\mathbf{H}) - \mathbf{H}\cdot(\nabla\times\mathbf{C})$, this is rearranged to produce:

$$\partial_t h + \frac{1}{2}\left[\sqrt{\frac{\epsilon_0}{\mu_0}}\nabla\cdot(\mathbf{E}\times\mathbf{A}) + \sqrt{\frac{\mu}{\epsilon}}\nabla\cdot(\mathbf{H}\times\mathbf{C})\right] = -\sqrt{\frac{\epsilon_0}{\mu_0}}\mathbf{E}\cdot\mathbf{B} + \sqrt{\frac{\mu_0}{\epsilon_0}}\mathbf{H}\cdot\mathbf{D}. \tag{12}$$

Inserting the relations $\mathbf{B} = \mu_0\mathbf{H}$ and $\mathbf{D} = \epsilon_0\mathbf{E}$ and using $\nabla(\mu_0/\epsilon_0) = 0$ leads to the free-space helicity continuity Equation [14,15,17]:

$$\partial_t h + \nabla\cdot\mathbf{v} = 0, \tag{13}$$

where the helicity flux density $\mathbf{v}$ is defined as:

$$\mathbf{v} = \frac{1}{2}\left[\sqrt{\frac{\epsilon_0}{\mu_0}}\mathbf{E}\times\mathbf{A} + \sqrt{\frac{\mu_0}{\epsilon_0}}\mathbf{H}\times\mathbf{C}\right]. \tag{14}$$

Thus, the conservation of helicity in a vacuum is explicitly demonstrated. Note that, thanks to the use of the definition (1), helicity is shown to be locally conserved; a stronger result than if only the integrated quantity were conserved. A pleasing analogy can be drawn between this characteristic of the helicity density and the one of the electromagnetic energy density: the total energy can be written as the volume integral of $\epsilon_0/2\left(\mathbf{E}^2 + c^2\mathbf{B}^2\right)$, or indeed as the integral over all space of either the $\mathbf{E}^2$ or $\mathbf{B}^2$ contributions. Only the energy density formed by the combined electric and magnetic contributions, however, is conserved locally [41].

There is an appealingly simple relationship between the helicity flux density and the spin density. The latter is often written as $\epsilon_0\mathbf{E}\times\mathbf{A}$, but it can also be written in the manifestly duplex-symmetric form:

$$\mathbf{s} = \frac{1}{2}\left[\epsilon_0\mathbf{E}\times\mathbf{A} + \mathbf{B}\times\mathbf{C}\right], \tag{15}$$

which immediately establishes the relation $\mathbf{s} = \mathbf{v}/c$. The relationship between these quantities and the helicity density is reminiscent of that between the energy density, the energy flux density and the momentum density, with Poynting's vector playing the role of both of the latter two quantities. For helicity and spin, $\mathbf{s}$ (or $\mathbf{v}$) plays an analogous double role.

3. Microscopic Sources

3.1. Helicity in the Presence of Current and Charge

We have introduced the helicity as a quantity associated with the free electromagnetic field, but it is also of interest as a way to describe the interaction of light with chiral or achiral matter. The presence of matter breaks dual-symmetry, as all known matter is made up of only electric charges, with no magnetic ones. In the presence of matter, therefore, helicity is not generally conserved. (We note parenthetically that the equations still remain invariant if we additionally "rotate" the electric charges into magnetic ones, introducing a charge and density and current density of magnetic charges. This symmetry of the equations means that it is in a sense a matter of convention that we speak of electric charges, rather than magnetic charges; the non-existence of magnetic monopoles can be rephrased as "every particle has the same ratio of electric to magnetic charge", and it is only a matter of convention that leads us to treat all charges as purely electric. See [41], Chapter 6 §11.).

One way of discussing helicity in the presence of charges is to treat the charges microscopically, with the fields described using the equations of free-space electromagnetism. In the presence of a current density $\mathbf{j}$ and a charge density ρ, it can be shown that the continuity equation becomes [17]:

$$\partial_t h + \nabla\cdot\mathbf{v} = \frac{1}{2}\sqrt{\frac{\mu_0}{\epsilon_0}}\left[\mathbf{g}\cdot(\nabla\times\mathbf{C}) + \mathbf{C}\cdot(\nabla\times\mathbf{g})\right], \tag{16}$$

where $\mathbf{g}$ is a vector field defined by the requirement $\nabla\times\mathbf{g} = \mathbf{j}^{\mathrm{T}}$, the transverse part of the current density. The continuity Equation (16) now expresses the non-conservation of helicity, with additional terms on the right-hand side showing how the matter acts as a source or sink of helicity in the field [17]. The terms "source" and "sink" must be treated with caution, as—unlike (for example) the case of energy—there is not a sense in which the matter "gains" or "loses" helicity when it absorbs or emits into the field.

The chirality density obeys a similar continuity equation [30]:

$$\partial_t\chi + \frac{1}{2\mu_0}\nabla\cdot\Big(\mathbf{E}\times(\nabla\times\mathbf{B}) - \mathbf{B}\times(\nabla\times\mathbf{E})\Big) = -\frac{1}{2}\left(\mathbf{j}\cdot(\nabla\times\mathbf{E}) + \mathbf{E}\cdot(\nabla\times\mathbf{j})\right). \tag{17}$$

Comparison of the two continuity equations again demonstrates the close connection between chirality and helicity: the chirality is the quantity one would obtain by forming the helicity from the curl of the fields, rather than the fields [15]. This observation makes clear why a direct proportionality holds for monochromatic light, where taking the curl merely introduces a factor of $i\omega$.

It is clear that $\mathbf{g}$ is acting like a magnetisation, as $\boldsymbol{\nabla} \times \mathbf{M} = \mathbf{j}_{\text{free}}$. We also note that, as we only consider the transverse $\mathbf{E}$ field, there is no analogous "polarisation-like" term to correspond with $\boldsymbol{\nabla} \cdot \mathbf{P} = \rho$. This leaves the source term asymmetric in terms of electric and magnetic contributions.

3.2. Dipole Model of a Helicity Source

As an illustration of how charges and currents can act as a source of helicity, we consider a point source consisting of an oscillating electric and a magnetic dipole, with electric and magnetic dipole moments $\mathbf{p}(t)$ and $\mathbf{m}(t)$, as has been examined by Leeder et al. [28]. This can be thought of as a simple model of a radiating chiral molecule, as the optical activity of chiral molecules ultimately arises from the simultaneous induction of electric and magnetic dipole moments through the electric-dipole magnetic-dipole polarisability tensor G [9] (see Chapter 3, §5.4). Leeder et al. treat the emission from the dipoles quantum-mechanically, calculating a differential irradiance for left- and right-circularly polarised light by considering the difference in decay rates into the two circular polarisation modes. From this, they obtain an expression for the net emitted helicity. There are three contributions to the total irradiance: one depending on $|\mathbf{p}|^2$, one on $|\mathbf{m}|^2$ and one on $\mathbf{p} \cdot \mathbf{m}$. The mixed dipole term $\mathbf{p} \cdot \mathbf{m}$ is the only contribution to the total irradiance that is different for the two enantiomers, and hence contributes to the net emitted helicity. We present here an analysis of helicity emission from the dipole system in the context of the continuity Equation (16), working within classical electromagnetism.

Consider an electric dipole oscillating along the $+z$ axis with dipole moment:

$$\mathbf{p}(t) = p_0 \exp[i(\omega t + \phi_{\text{p}})]\hat{\mathbf{z}}, \tag{18}$$

where we have defined $p_0 = q_0 d$, with d and q_0 the size and charge of the dipole, respectively. The resultant (retarded) vector potential in the far field defined by $d \ll \lambda \ll r$ is [43]:

$$\mathbf{A}'_{\text{p}} = \Re\left[\frac{i\omega\mu_0 p_0}{4\pi r} \exp[i(\omega(t - r/c) + \phi_{\text{p}})]\hat{\mathbf{z}}\right]. \tag{19}$$

We call this $\mathbf{A}'_{\text{p}}$, as we reserve the symbol $\mathbf{A}$ for the transverse part of the vector potential. From this expression, $\mathbf{B}_p$, $\mathbf{E}_{\text{p}}$, $\mathbf{C}_{\text{p}}$ and $\mathbf{A}_{\text{p}} \equiv \mathbf{A}'^{\text{T}}_{\text{p}}$ can be found.

Similarly, we consider a current oscillating in a loop of radius b in the xy plane, $I(t) = I_0 \exp[i(\omega t + \phi_{\text{m}})]$. The resultant oscillating dipole moment is given by:

$$\mathbf{m}(t) = \int I(t)\, d\mathbf{a} = m_0 \exp[i(\omega t + \phi_{\text{m}})]\hat{\mathbf{z}}, \tag{20}$$

where $m_0 = \pi b^2 I_0$. The vector potential is calculated as [43]:

$$\mathbf{A}_{\text{m}} = \mathbf{A}^{\text{T}}_{\text{m}} = \Re\left[\frac{i\omega\mu_0 m_0}{4\pi r^2 c}\,(x\hat{\mathbf{y}} - y\hat{\mathbf{x}})\right]. \tag{21}$$

Using this to find $\mathbf{B}_{\text{m}}$, $\mathbf{E}_{\text{m}}$ and $\mathbf{C}_{\text{m}}$, we calculate the total helicity density and flux density of the combined electric-magnetic dipole system in the far-field:

$$h_{\text{pm}} = \frac{\epsilon_0}{c}\left(\frac{\mu_0^2 m_0 p_0 \omega^3}{16\pi^2 r^4}\left(x^2 + y^2\right)\sin(\phi_{\text{m}} - \phi_{\text{p}})\right), \tag{22}$$

$$\mathbf{v}_{\mathrm{pm}} = \epsilon_0 \left(\frac{\mu_0^2 m_0 p_0 \omega^3}{16\pi^2 r^5} \left(x^2 + y^2\right) \sin(\phi_{\mathrm{m}} - \phi_{\mathrm{p}}) \right) \mathbf{r}. \tag{23}$$

From (22), $\partial_t h_{\mathrm{pm}} = 0$, and we use Gauss's theorem to calculate the net helicity flux of the combined dipole system:

$$\int \boldsymbol{\nabla} \cdot \mathbf{v}_{\mathrm{pm}} \, d^3\mathbf{r} = \frac{\epsilon_0 \mu_0^2 m_0 p_0 \omega^3}{6\pi} \sin(\phi_{\mathrm{m}} - \phi_{\mathrm{p}}). \tag{24}$$

We can identify this with the source term on the right-hand side of (16). Noting $\int \mathbf{g} \cdot (\boldsymbol{\nabla} \times \mathbf{C}) \, d^3\mathbf{r} = \int \mathbf{C} \cdot (\boldsymbol{\nabla} \times \mathbf{g}) \, d^3\mathbf{r}$ [17], we obtain:

$$\int \mathbf{g} \cdot (\boldsymbol{\nabla} \times \mathbf{C}) \, d^3\mathbf{r} = \frac{\epsilon_0 \mu_0 m_0 p_0 \omega^3}{6\pi c} \sin(\phi_{\mathrm{m}} - \phi_{\mathrm{p}}), \tag{25}$$

which is maximised for a phase difference of $\pm\pi/2$ between the dipoles.

Left- and right-circularly polarised light can also be produced by a pair of orthogonally-aligned oscillating electric or magnetic dipoles. We consider the former configuration: $\mathbf{p}_1(t) = |\mathbf{p}_1(t)|\hat{\mathbf{z}}$ and $\mathbf{p}_2(t) = |\mathbf{p}_2(t)|\hat{\mathbf{y}}$, with a phase difference of $\phi_{\mathrm{p}_2} - \phi_{\mathrm{p}_1}$ between the oscillations. The resultant helicity flux density in the far-field is:

$$\mathbf{v}_{\mathrm{pp}} = \frac{\epsilon_0 c}{2} \left(\frac{\mu_0^2 p_0^2 \omega^3}{16\pi^2 r^5} \sin(\phi_{\mathrm{p}_2} - \phi_{\mathrm{p}_1}) x r \right) \mathbf{r}, \tag{26}$$

again maximised for $\phi_{\mathrm{p}_2} - \phi_{\mathrm{p}_1} = \pi/2$. The helicity density of this coupled dipole system is again time independent, so that the "source term" is found in analogy with the above calculation. We find $\int \boldsymbol{\nabla} \cdot \mathbf{v}_{\mathrm{pp}} \, d^3\mathbf{r}=0$: the coupled electric-electric dipole system does not describe a source of helicity.

These results follow intuitively by considering Figures 2 and 3, where the far-field patterns of the electric-magnetic and electric-electric dipole systems in the yz plane have been drawn. A $\pi/2$ phase difference between the dipole oscillations results in right-circular polarisation in the $+x$ direction in both dipole systems. In the $-x$ direction, however, we obtain left-circular polarisation for the electric-magnetic dipole pair and right-circular polarisation for the electric-electric dipoles. Thus, we obtain a net flux of helicity for the electric-magnetic dipole system only.

The results of this calculation are intimately related to the respective parity of the electric-magnetic and electric-electric dipole systems. Considering again Figure 2, a parity transformation reverses the direction of the electric dipole only, so that the field patterns in both the $+x$ and the $-x$ direction are reversed, and a net flux of left-circular polarisation is produced. A parity transformation $\hat{\mathbf{P}}$ thus interconverts the two "enantiomeric" configurations of the electric-magnetic dipole systems and produces a negative sign in the integrated source term in (25). The two enantiomeric forms can therefore be described simply by the aligned and anti-aligned configurations (or equivalently, as a positive or negative phase difference between the two oscillations in either configuration), which can be distinguished in the far-field yz plane by an excess of right- or left-circular polarisation, respectively.

For the electric-electric dipoles in Figure 3, however, both dipoles are reversed under $\hat{\mathbf{P}}$, so that the field patterns in $\pm x$ directions remain unchanged. As the electric-electric system is parity-even, but left- and right-circularly polarised fields are reversed under a parity transformation, it follows that the electric-electric dipole system cannot produce an excess of either polarisation and hence cannot be used to describe a source of helicity.

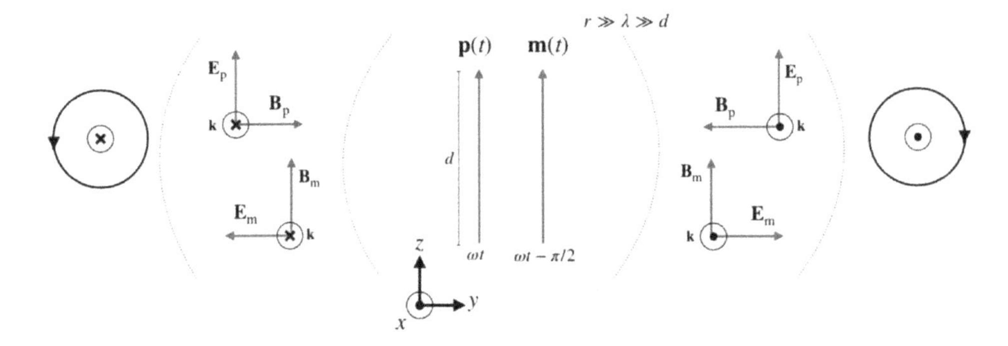

Figure 2. Oscillating electric $\mathbf{p}(t)$ and magnetic $\mathbf{m}(t)$ dipoles aligned along the z axis, with the latter lagging by a phase of $\pi/2$. The far-field pattern is that of right-circularly polarised light in the $+x$ direction and left-circularly polarised light in the $-x$ direction: there is a net flux of right-circular polarisation in the yz plane.

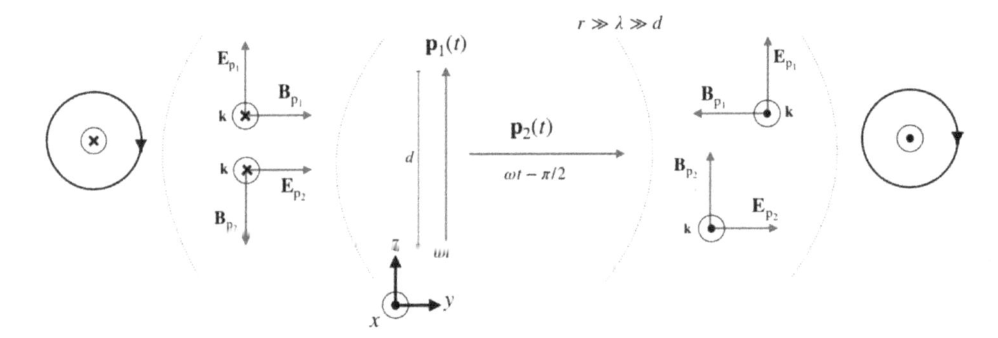

Figure 3. Two oscillating electric dipoles, labelled $\mathbf{p}_1(t)$ and $\mathbf{p}_2(t)$, oscillate along the z and y axes, respectively, with a phase difference of $\pi/2$. The far-field pattern in the $+x$ direction is identical to that produced by the electric-magnetic dipole configuration in Figure 2, but has opposite polarisation in the $-x$ direction: the net helicity flux in the yz plane is zero.

4. Macroscopic Sources

4.1. Helicity in Achiral, Reciprocal Media

We have seen how helicity can be produced by a microscopic source (a dipole), but it is also possible to discuss the generation of helicity within the framework of macroscopic electromagnetism in dielectric media. It is perhaps surprising that even in the presence of matter, there are some situations in which the helicity of an arbitrary electromagnetic field is still conserved. The conditions under which the electromagnetic helicity is conserved within media have been studied in recent years [16,18,19]. Fernandez-Corbaton et al. [16] consider the propagation of helicity in isotropic, lossless linear media, and these results are extended by van Kruining and Götte in [18] to include anisotropic and general linear media. Alpeggiani et al. [19] further consider helicity in a dispersive, lossy medium, while the electromagnetic chirality is examined in such media by Vázquez-Lozano and Martínez [44].

In the following, we consider a linear, lossless and isotropic medium. If the medium is comprised of distinct, homogeneous regions labelled by i, helicity is conserved so long as ϵ_i/μ_i remains constant for all i [16]. Following a similar method to that used in Section 2.5, we can derive a continuity equation for helicity in such media. We use the definitions of h (1) and $\mathbf{v}$ (14), along with the constitutive relations $\mathbf{D} = \epsilon\mathbf{E}$ and $\mathbf{B} = \mu\mathbf{H}$, to obtain:

$$\partial_t h + \boldsymbol{\nabla} \cdot \mathbf{v} = \frac{1}{2}\left[\boldsymbol{\nabla}\left(\sqrt{\frac{\epsilon}{\mu}}\right) \cdot \left(\mathbf{E} \times \mathbf{A}\right) + \boldsymbol{\nabla}\left(\sqrt{\frac{\mu}{\epsilon}}\right) \cdot \left(\mathbf{H} \times \mathbf{C}\right)\right], \tag{27}$$

where we have allowed for the possibility that ϵ and μ are functions of position. This explicitly shows the conservation of helicity when $\boldsymbol{\nabla}\,(\epsilon/\mu) = 0$, which can be seen as a continuous statement of the result for stratified media presented in [16].

We extend the results of [17] discussed in Section 3.1, to examine the effects of inserting a local current density $\mathbf{j}$ into a medium described by the constitutive relations $\mathbf{D} = \epsilon\mathbf{E}$ and $\mathbf{B} = \mu\mathbf{H}$. We find a source term analogous to the right-hand side of (16), but with the replacement of $\epsilon_0 \to \epsilon$ and $\mu_0 \to \mu$. Moreover, this type of source cannot be associated with that produced by a gradient of ϵ/μ, as given by the right-hand side of (27).

4.2. Helicity in Bi-Isotropic Media

We model a general linear, lossless bi-isotropic medium, where we allow for both a chiral and a magnetoelectric response, using an extension of the "Drude–Born–Fedorov" constitutive relations to include the Tellegen parameter α [45]:

$$\begin{aligned} \mathbf{D} &= \epsilon\,(\mathbf{E} + \beta\boldsymbol{\nabla} \times \mathbf{E}) + \alpha\mathbf{H}, \\ \mathbf{B} &= \mu\,(\mathbf{H} + \beta\boldsymbol{\nabla} \times \mathbf{H}) + \alpha\mathbf{E}, \end{aligned} \tag{28}$$

where β is the referred to as the chirality parameter. The results of [16] have been extended to include chiral and Tellegen media, as well an anisotropic polarisabilities, in [18], where it has been explicitly shown that Maxwell's equations in a medium remain invariant under a duality transformation of the fields when there is a constant ratio ϵ/μ and a Tellegen parameter of zero. The chirality parameter, on the other hand, is free to vary in space. This result in [18] is based on the symmetrized constitutive relations of Condon [45], but it is straightforward to show that the same condition for dual symmetry holds for the Drude–Born–Fedorov relations given above.

Inserting the constitutive relations (28) with $\alpha = 0$ into the helicity density and flux density (1) and (14) reveals that the conservation of helicity in a chiral medium cannot be expressed by a local continuity equation unless the expression for the helicity density inside a chiral medium is suitably modified [46].

Helicity Conservation in a Chiral Medium

Both helicity and energy are conserved in a lossless chiral medium [18,47]. It therefore follows that, in a situation where the interface between the vacuum and the medium is dual-symmetric, the helicity per photon of light in the chiral medium should remain the same as the vacuum values. This is the situation depicted in Figure 4. In [14], free electromagnetic fields with left- and right-circular polarisation are shown to have a helicity of $\pm\hbar$ per photon. In [46], the expressions for the helicity density and flux density (1) and (14) are trivially extended to those in a linear medium by replacement of the vacuum electric and magnetic responses, ϵ_0 and μ_0, with ϵ and μ. In considering the propagation of left- and right-circularly polarised light within a chiral medium, is it shown that, although the flux density formed in this way produces the correct helicity of $\pm\hbar$ per photon, the helicity density does not. This follows from the fact that the energy density within a chiral medium is of the form [48]:

$$w_1 = \frac{1}{2}\left[\mathbf{D} \cdot \mathbf{E} + \mathbf{B} \cdot \mathbf{H} - \beta\epsilon\mu(\mathbf{E} \cdot \dot{\mathbf{H}} - \dot{\mathbf{E}} \cdot \mathbf{H})\right], \tag{29}$$

containing an explicit β-dependent term. As we know the correct form of the helicity flux density within the medium is:

$$\mathbf{v} = \frac{1}{2}\left(\sqrt{\frac{\epsilon}{\mu}}\mathbf{E}\times\mathbf{A} + \sqrt{\frac{\mu}{\epsilon}}\mathbf{H}\times\mathbf{C}\right), \tag{30}$$

we can use this and the condition of local helicity conservation to find the form of the helicity density. We find:

$$\begin{aligned}\nabla\cdot\mathbf{v} &= \frac{1}{2}\left(\sqrt{\frac{\epsilon}{\mu}}\left(\mathbf{A}\cdot(\nabla\times\mathbf{E}) - \mathbf{E}\cdot(\nabla\times\mathbf{A})\right) - \sqrt{\frac{\mu}{\epsilon}}\left(\mathbf{C}\cdot(\nabla\times\mathbf{H}) - \mathbf{H}\cdot(\nabla\times\mathbf{C})\right)\right)\\ &= \frac{1}{2}\left(\sqrt{\frac{\epsilon}{\mu}}\left(-\mathbf{A}\cdot\dot{\mathbf{B}} - \mathbf{E}\cdot\mathbf{B}\right) - \sqrt{\frac{\mu}{\epsilon}}\left(\mathbf{C}\cdot\dot{\mathbf{D}} + \mathbf{H}\cdot\mathbf{D}\right)\right),\end{aligned} \tag{31}$$

and use the product rule to write this as:

$$\nabla\cdot\mathbf{v} = -\partial_t\frac{1}{2}\left(\sqrt{\frac{\epsilon}{\mu}}\mathbf{A}\cdot\mathbf{B} - \sqrt{\frac{\mu}{\epsilon}}\mathbf{C}\cdot\mathbf{D}\right) - \left(\sqrt{\frac{\epsilon}{\mu}}\mathbf{E}\cdot\mathbf{B} - \sqrt{\frac{\mu}{\epsilon}}\mathbf{H}\cdot\mathbf{D}\right). \tag{32}$$

Identifying the time derivative as $\partial_t h$ and inserting the constitutive relations (28) with $\alpha = 0$ leads to:

$$\nabla\cdot\mathbf{v} = -\partial_t h - \sqrt{\epsilon\mu}\beta\left[\mathbf{E}\cdot\dot{\mathbf{D}} + \mathbf{H}\cdot\dot{\mathbf{B}}\right], \tag{33}$$

which is rearranged to produce:

$$\partial_t\left(h + \sqrt{\epsilon\mu}\beta w\right) + \nabla\cdot\mathbf{v} = 0, \tag{34}$$

where $w = 1/2\left(\mathbf{D}\cdot\mathbf{E} + \mathbf{B}\cdot\mathbf{H}\right)$ is the energy density in an achiral medium. Equation (34) is correct to first order in the chirality parameter, denoted $\mathcal{O}(\beta)$, by which we mean that we neglect terms multiplied by β^2, or higher powers. This is incorporated into the definition of the helicity density h to form [46]:

$$h_1 = \frac{1}{2}\left(\sqrt{\frac{\epsilon}{\mu}}\mathbf{A}\cdot\mathbf{B} - \sqrt{\frac{\mu}{\epsilon}}\mathbf{C}\cdot\mathbf{D}\right) + \sqrt{\epsilon\mu}\beta w. \tag{35}$$

It is further shown in [46] that this indeed produces a helicity to energy density ratio of $\pm 1/\omega$ for right- and left-circularly polarised light, leading to a helicity of $\pm\hbar$ per photon [47]. In addition, it can be shown that higher order terms in β of the helicity and energy densities retain this correspondence between the definitions, i.e.:

$$h_{n+1} = \frac{1}{2}\left(\sqrt{\frac{\epsilon}{\mu}}\mathbf{A}\cdot\mathbf{B} - \sqrt{\frac{\mu}{\epsilon}}\mathbf{C}\cdot\mathbf{D}\right) + \sqrt{\epsilon\mu}\beta w_n, \tag{36}$$

such that $\partial_t w_n + \nabla\cdot\mathbf{S} = 0$ and $\partial_t h_{n+1} + \nabla\cdot\mathbf{v} = 0$, where w_n is the energy density of the fields to $\mathcal{O}(\beta^n)$ and $\mathbf{S} = \mathbf{E}\times\mathbf{H}$ is the energy flux density. This follows from the exact expression:

$$\partial_t h + \nabla\cdot\mathbf{v} = \sqrt{\epsilon\mu}\beta\nabla\cdot\mathbf{S}, \tag{37}$$

such that energy conservation to all orders in β implies helicity conservation to all orders in β.

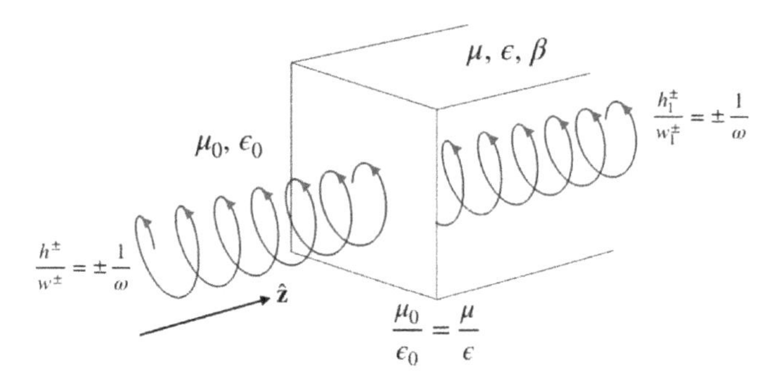

Figure 4. At the interface between a vacuum and a dual-symmetric, lossless chiral medium characterised by ϵ, μ and the chirality parameter β, both the energy and helicity of an electromagnetic field are conserved. As a consequence, the ratio of the helicity density to energy density, h/w, must be preserved across the interface. In the chiral medium, the energy density contains a chiral contribution, as given by w_1 (29), requiring a modification of the helicity density to h_1 (35) such that $h/w = h_1/w_1$ holds.

We can examine the general form of a source of helicity in lossless non-reciprocal media by inserting the constitutive relations (28) with $\alpha \neq 0$ into h_1 and $\mathbf{v}$, producing:

$$\partial_t h_1 + \nabla \cdot \mathbf{v} = -\alpha \left(\sqrt{\frac{\epsilon}{\mu}} |\mathbf{E}|^2 - \sqrt{\frac{\mu}{\epsilon}} |\mathbf{H}|^2 \right), \tag{38}$$

where we have imposed $\nabla(\epsilon/\mu) = 0$. As the energy is conserved [47], it follows that the helicity per photon of the light within a Tellegen material must differ from the familiar free-space values.

4.3. Currents and Charges in Bi-Isotropic Media

We now consider what happens when a local current density is placed inside a chiral medium. Inserting the constitutive relations (28) into $\nabla \times \mathbf{E} = -\dot{\mathbf{B}}$ and $\nabla \times \mathbf{H} = \dot{\mathbf{D}} + \mathbf{j}$ leads to:

$$\dot{\mathbf{A}} = \beta (\nabla \times \mathbf{E}) + \frac{1}{\epsilon} (\nabla \times \mathbf{C} + \alpha \mathbf{H}) \tag{39}$$

and:

$$\dot{\mathbf{C}} = \mathbf{g} + \beta (\nabla \times \mathbf{H}) - \frac{1}{\mu} (\nabla \times \mathbf{A} - \alpha \mathbf{E}), \tag{40}$$

where again, $\nabla \times \mathbf{g}$ is equal to the transverse part of the current density. We insert $\dot{\mathbf{A}}$ and $\dot{\mathbf{C}}$ into $\partial_t h_1$ from (35) and find to $\mathcal{O}(\beta)$:

$$\partial_t h_1 + \nabla \cdot \mathbf{v} = \frac{1}{2}\sqrt{\frac{\mu}{\epsilon}} [\mathbf{g} \cdot (\nabla \times \mathbf{C}) + \mathbf{C} \cdot (\nabla \times \mathbf{g})] - \sqrt{\epsilon\mu}\beta \mathbf{j} \cdot \mathbf{E} - \alpha \left(\sqrt{\frac{\epsilon}{\mu}} |\mathbf{E}|^2 - \sqrt{\frac{\mu}{\epsilon}} |\mathbf{H}|^2 \right). \tag{41}$$

This reduces to the result for a medium with no chiroptical or magnetoelectric response when $\alpha = \beta = 0$. The helicity contribution due to the chiral response of the material is identifiable as an energy source, as obtained from the energy continuity equation in the presence of charges [41]. As this term is a scalar, the pseudoscalar nature of β is responsible for this term's acting as a source of helicity. The Tellegen contribution, proportional to α, is identical to the helicity source in an absence of currents or charges, as given in (38).

Consider a chiral, reciprocal material described by (28) with $\alpha = 0$. From (41), we would expect that an emitter that in free space emits no net helicity, such as a single oscillating electric dipole, may act as a source of helicity when placed inside a chiral medium. Lathakia et al. have shown that this is so, by explicitly calculating the radiation pattern of a single oscillating electric dipole embedded in a sphere composed of a lossless chiral medium [49]. From the point of view of an observer in the far field, outside the sphere, the radiation pattern from the point electric dipole embedded in the sphere appears identical to that of a point electric and point magnetic dipole oscillating in a vacuum. In particular, the chiral medium is impedance matched with the surrounding vacuum to produce this result, so that the net helicity cannot be attributed to a gradient in ϵ/μ at the vacuum-chiral interface. As the chiral medium considered above and in [49] is lossless, neither can this generation of helicity arise as a result of circular dichroism within the chiral sphere [30]. Furthermore, we know that the electromagnetic helicity of a field is conserved within a dual-symmetric chiral medium, so the helicity "source" in this case can only be attributed to the interaction of the embedded current and the chiral medium itself; a result which seems worthy of further investigation.

It is interesting to observe that for $\mathcal{O}(\beta)$, we can write $-\beta\mathbf{j}\cdot\mathbf{E} = \beta\left(\nabla\times\mathbf{g}\right)\cdot\left(\nabla\times\mathbf{C}\right)$, so that the current helicity source in the chiral medium appears as an even-parity combination of $\mathbf{g}$ and $\mathbf{C}$. We would expect terms $\mathcal{O}(\beta^{2n+1})$ to echo this structure, with even-order terms $\mathcal{O}(\beta^{2n})$ containing odd-parity combinations of $\mathbf{g}$ and $\mathbf{C}$, such as given by the first term on the right-hand side of (41).

5. Concluding Remarks

We have examined in detail the construction of the helicity density (1) in a vacuum and discussed the merits of using this quantity to characterise chiral light. Alongside this, we examined the chirality density (2) of the free electromagnetic field and pointed out cases in which the two quantities are trivially related. Only the conservation of helicity, however, generates a physically-meaningful symmetry transformation of the system. We used this to construct a continuity equation of the helicity of the free field, before extending the method to examine cases under which this symmetry is broken and helicity is no longer conserved. We have identified four distinct types of helicity sources. The first results from a non-constant value of ϵ/μ, taking the form of the right-hand side of (27). The second helicity source has the general form of (16) [17] and can be understood in its simplest form in terms of a coupled electric-magnetic dipole system. The third results from the non-reciprocity parameter α in the constitutive relations, of the form (38). The final type of source examined in this article results from a dual-symmetric object embedded in a chiral medium, expressible as an energy source multiplied by the chirality parameter β, as given by (41).

Categorisation of the distinct sources of helicity in this way provides insight into the electromagnetic response of different types of matter and is achieved by exploiting the inherent symmetry of the Maxwell equations. The distinction between the microscopic and macroscopic sources results from the non-tractable nature of the problem of determining the electromagnetic response of large volumes ($\gg$than the size of individual molecules) of helicity sources. In order to bridge this gap, we can perhaps look to experiments in which chiral objects are embedded into a dielectric host, forming an artificial composite chiral medium [50,51]. In recent work, such methods have been used to verify that a chiral nanostructure is in fact able to sense the orbital, as well as the spin angular momentum of an impinging light beam [52,53]. Theoretical study on this subject continues to reveal insight into the significance of twisted light beams in chiral light–matter interactions [54,55], paving the way for new methods in the detection and manipulation of chiral matter. The importance of helicity in the characterisation of both natural and engineered chiral nanostructures is indeed becoming increasingly apparent: Hanifeh et al. [56] show that using structured light with maximised helicity leads to a direct measure of the chirality of such an object, which does not require knowledge of the helicity or energy densities of the field. It is also evident that both the helicity of the incident fields and the dual-symmetry (or helicity preserving nature) of a photonic structure are essential in circular dichroism enhancement effects [57]. Reconciling our understanding of both microscopic and macroscopic sources of helicity in a general theoretical model, however, is an ongoing topic of investigation.

Author Contributions: This paper was built on original ideas by each of the authors, and each author performed some of the calculations. All authors contributed to the writing and editing of the manuscript.

Funding: This research was funded by The Royal Society under Grant Numbers RP/EA/180010 and RP/150122 and the Engineering and Physical Sciences Research Council under Grant Number EP/N509668/1.

Conflicts of Interest: The authors declare no conflicts of interest.

References

1. Arago, D. Sur une modification remarquable qu'éprouvent les rayons lumineux dans leur passage à travers certains corps diaphanes, et sur quelques autres nouveaux phénomènes d'optique. *Mem. Inst.* **1811**, *1*, 93–134.
2. Biot, J.B. Phénomènes de polarisation successive, observés dans les fluides homogènes. *Bull. Soc. Philomath.* **1815**, *1*, 190–192.
3. Pasteur, L. Recherches sur les relations qui peuvent exister entre la forme cristalline, la composition chimique et le sens de la polarisation rotatoire. *Ann. Chim. Phys.* **1848**, *24*, 442–459.
4. Van'T Hoff, J.H.; Pasteur, L.; Richardson, G.M.; Le Bel, J.A. *The Foundations of Stereo Chemistry: Memoirs by Pasteur, Van'T Hoff, Lebel and Wilslincenus*; American Book Company: Woodstock, GA, USA, 1901.
5. Kelvin, L. *The Molecular Tactics of a Crystal*; Clarendon Press: Oxford, UK, 1894.
6. Cameron, R.P.; Götte, J.B.; Barnett, S.M.; Yao, A.M. Chirality and the angular momentum of light. *Philos. Trans. R. Soc. A Math. Phys. Eng. Sci.* **2017**, *375*, 20150433. [CrossRef] [PubMed]
7. Lee, T.D.; Yang, C.N. Question of parity conservation in weak interactions. *Phys. Rev.* **1956**, *104*, 254–258. [CrossRef]
8. Wagnière, G.H. *On Chirality and the Universal Asymmetry: Reflections on Image and Mirror Image*; John Wiley and Sons: Hoboken, NJ, USA, 2007.
9. Barron, L.D. *Molecular Light Scattering and Optical Activity*; Cambridge University Press: Cambridge, UK, 2004.
10. Cotton, A. Absorption inégale des rayons circulaires droit et gauche dans certains corps actifs. *Compt. Rend.* **1895**, *120*, 989–991.
11. Poynting, J.H. The Wave Motion of a Revolving Shaft, and a Suggestion as to the Angular Momentum in a Beam of Circularly Polarised Light. *Proc. R. Soc. A Math. Phys. Eng. Sci.* **1909**, *82*, 560–567. [CrossRef]
12. Calkin, M.G. An Invariance Property of the Free Electromagnetic Field. *Am. J. Phys.* **1965**, *33*, 958–960. [CrossRef]
13. Trueba, J.L.; Rañada, A.F. The electromagnetic helicity. *Eur. J. Phys.* **1996**, *17*, 141–144. [CrossRef]
14. Barnett, S.M.; Cameron, R.P.; Yao, A.M. Duplex symmetry and its relation to the conservation of optical helicity. *Phys. Rev. A* **2012**, *86*, 013845. [CrossRef]
15. Cameron, R.P.; Barnett, S.M.; Yao, A.M. Optical helicity, optical spin and related quantities in electromagnetic theory. *New J. Phys.* **2012**, *14*, 053050. [CrossRef]
16. Fernandez-Corbaton, I.; Zambrana-Puyalto, X.; Tischler, N.; Vidal, X.; Juan, M.L.; Molina-Terriza, G. Electromagnetic duality symmetry and helicity conservation for the macroscopic maxwell's equations. *Phys. Rev. Lett.* **2013**, *111*, 060401. [CrossRef] [PubMed]
17. Nienhuis, G. Conservation laws and symmetry transformations of the electromagnetic field with sources. *Phys. Rev. A* **2016**, *93*, 023840. [CrossRef]
18. van Kruining, K.; Götte, J.B. The conditions for the preservation of duality symmetry in a linear medium. *J. Opt.* **2016**, *18*, 085601. [CrossRef]
19. Alpeggiani, F.; Bliokh, K.Y.; Nori, F.; Kuipers, L. Electromagnetic Helicity in Complex Media. *Phys. Rev. Lett.* **2018**, *120*, 243605. [CrossRef] [PubMed]
20. Saffman, P.G. *Vortex Dynamics*; Cambridge University Press: New York, NY, USA, 1992.
21. Madja, A.J.; Bertozzi, A.L. *Vorticity and Incompressible Flow*; Cambridge University Press: Cambridge, UK, 2002.
22. Moffatt, H.K. The degree of knottedness of tangled vortex lines. *J. Fluid Mech.* **1969**, *35*, 117–129. [CrossRef]
23. Woltjer, L. A theorem on force-free magnetic fields. *Proc. Natl. Acad. Sci. USA* **1958**, *44*, 489–491. [CrossRef] [PubMed]

24. Priest, E.; Forbes, T. *Magnetic Reconnection: MHD Theory and Applications*; Cambridge University Press: New York, NY, USA, 2000.
25. Cameron, R.P. On the 'second potential' in electrodynamics. *J. Opt.* **2014**, *16*, 015708. [CrossRef]
26. Fernandez-Corbaton, I.; Vidal, X.; Tischler, N.; Molina-Terriza, G. Necessary symmetry conditions for the rotation of light. *J. Chem. Phys.* **2013**, *138*, 214311. [CrossRef] [PubMed]
27. Coles, M.M.; Andrews, D.L. Chirality and angular momentum in optical radiation. *Phys. Rev. A* **2012**, *85*, 63810. [CrossRef]
28. Leeder, J.M.; Haniewicz, H.T.; Andrews, D.L. Point source generation of chiral fields: Measures of near-and far-field optical helicity. *J. Opt. Soc. Am. B* **2015**, *32*, 2308–2313. [CrossRef]
29. Lipkin, D.M. Existence of a new conservation law in electromagnetic theory. *J. Math. Phys.* **1964**, *5*, 696–700. [CrossRef]
30. Tang, Y.; Cohen, A.E. Optical Chirality and Its Interaction with Matter. *Phys. Rev. Lett.* **2010**, *104*, 163901. [CrossRef] [PubMed]
31. Tang, Y.; Cohen, A.E. Enhanced Enantioselectivity in Excitation of Chiral Molecules by Superchiral Light. *Science* **2011**, *332*, 333–336. [CrossRef] [PubMed]
32. Abdulrahman, N.; Syme, C.D.; Jack, C.; Karimullah, A.; Barron, L.D.; Gadegaard, N.; Kadodwala, M. The origin of off-resonance non-linear optical activity of a gold chiral nanomaterial. *Nanoscale* **2013**, *5*, 12651–12657. [CrossRef] [PubMed]
33. Andrews, D.L.; Coles, M.M. Optical superchirality and electromagnetic angular momentum. In *Complex Light and Optical Forces VI*; International Society for Optics and Photonics: Bellingham, WA, USA, 2012; Volume 8274, pp. 827405–827407.
34. van Kruining, K.C.; Cameron, R.P.; Götte, J.B. Superpositions of up to six plane waves without electric-field interference. *Optica* **2018**, *5*, 1091–1098. [CrossRef]
35. Karczmarek, J.; Wright, J.; Corkum, P.; Ivanov, M. Optical centrifuge for molecules. *Phys. Rev. Lett.* **1999**, *82*, 3420–3423. [CrossRef]
36. Van Enk, S.J.; Nienhuis, G. Spin and Orbital Angular Momentum of Photons. *Europhys. Lett.* **1994**, *25*, 497–501. [CrossRef]
37. van Enk, S.J.; Nienhuis, G. Commutation rules and eigenvalues of spin and orbital angular momentum of radiation fields. *J. Mod. Opt.* **1994**, *41*, 963–977. [CrossRef]
38. Barnett, S.M. Rotation of electromagnetic fields and the nature of optical angular momentum. *J. Mod. Opt.* **2010**, *57*, 1339–1343. [CrossRef] [PubMed]
39. Barnett, S.M.; Allen, L.; Cameron, R.P.; Gilson, C.R.; Padgett, M.J.; Speirits, F.C.; Yao, A.M. On the natures of the spin and orbital parts of optical angular momentum. *J. Opt.* **2016**, *18*, 064004. [CrossRef]
40. Cameron, R.P.; Speirits, F.C.; Gilson, C.R.; Allen, L.; Barnett, S.M. The azimuthal component of Poynting's vector and the angular momentum of light. *J. Opt.* **2015**, *17*, 125610–125618. [CrossRef]
41. Jackson, J.D. *Classical Electrodynamics*, 3rd ed.; Wiley: New York, NY, USA, 1962.
42. Berry, M.V. Optical currents. *J. Opt. A Pure Appl. Opt.* **2009**, *11*, 094001. [CrossRef]
43. Griffiths, D.J. *Introduction to Electrodynamics*, 3rd ed.; Prentice Hall: Upper Saddle River, NJ, USA, 1999.
44. Vázquez-Lozano, J.E.; Martínez, A. Optical Chirality in Dispersive and Lossy Media. *Phys. Rev. Lett.* **2018**, *121*, 043901. [CrossRef] [PubMed]
45. Lakhtakia, A. *Beltrami Fields in Chiral Media*; World Scientific Series in Contemporary Chemical Physics; World Scientific: Singapore, 1994.
46. Crimin, F.; Mackinnon, N.; Götte, J.B.; Barnett, S.M. On the helicity density in a chiral medium. **2019**, in press.
47. Barnett, S.M.; Cameron, R.P. Energy conservation and the constitutive relations in chiral and non-reciprocal media. *J. Opt.* **2016**, *18*, 015404. [CrossRef]
48. Bursian, V.; Timorew, F.A. Zur Theorie der optisch aktiven isotropen Medien. *Z. Phys.* **1926**, *38*, 475–484. [CrossRef]
49. Lakhtakia, A.; Varadan, V.K.; Varadan, V.V. Radiation by a point electric dipole embedded in a chiral sphere. *J. Phys. D App. Phys.* **1990**, *23*, 481–485. [CrossRef]
50. Elezzabi, A.Y.; Sederberg, S. Optical activity in an artificial chiral media: A terahertz time-domain investigation of Karl F Lindman's 1920 pioneering experiment. *Opt. Express* **2009**, *17*, 6600–6612. [CrossRef] [PubMed]

51. Wang, Z.; Cheng, F.; Winsor, T.; Liu, Y. Optical chiral metamaterials: A review of the fundamentals, fabrication methods and applications. *Nanotechnology* **2016**, *27*, 412001–412021. [CrossRef] [PubMed]
52. Brullot, W.; Vanbel, M.K.; Swusten, T.; Verbiest, T. Resolving enantiomers using the optical angular momentum of twisted light. *Sci. Adv.* **2016**, *2*, e1501349. [CrossRef] [PubMed]
53. Woźniak, P.; De León, I.; Höflich, K.; Leuchs, G.; Banzer, P. Interaction of light carrying orbital angular momentum with a chiral dipolar scatterer. *arXiv* **2019**, arXiv:1902.01731.
54. Forbes, K.A.; Andrews, D.L. Optical orbital angular momentum: Twisted light and chirality. *Opt. Lett.* **2018**, *43*, 435–438. [CrossRef] [PubMed]
55. Forbes, K.A.; Andrews, D.L. The angular momentum of twisted light in anisotropic media: Chiroptical interactions in chiral and achiral materials. In *Nanophotonics VII*; International Society for Optics and Photonics: Bellingham, WA, USA, 2018; Volume 10672.
56. Hanifeh, M.; Albooyeh, M.; Capolino, F. Optimally Chiral Electromagnetic Fields: Helicity Density and Interaction of Structured Light with Nanoscale Matter. *arXiv* **2018**, arXiv:1809.04117.
57. Graf, F.; Feis, J.; Garcia-Santiago, X.; Wegener, M.; Rockstuhl, C.; Fernandez-Corbaton, I. Achiral, Helicity Preserving, and Resonant Structures for Enhanced Sensing of Chiral Molecules. *ACS Photonics* **2019**. [CrossRef]

Review

Generation of Orbital Angular Momentum Modes Using Fiber Systems

Hongwei Zhang, Baiwei Mao, Ya Han, Zhi Wang, Yang Yue and Yange Liu *

Tianjin Key Laboratory of Optoelectronic Sensor and Sensing Network Technology, Institute of Modern Optics, Nankai University, Tianjin 300350, China; 2120150223@mail.nankai.edu.cn (H.Z.); maobaiwei@mail.nankai.edu.cn (B.M.); hany1020@163.com (Y.H.); zhiwang@nankai.edu.cn (Z.W.); yueyang@nankai.edu.cn (Y.Y.)

* Correspondence: ygliu@nankai.edu.cn

Received: 30 January 2019; Accepted: 8 March 2019; Published: 12 March 2019

Featured Application: In this paper, the basic concepts of fiber modes, the principle of generation and detection of orbital angular momentum (OAM) modes are exhaustive discussed, and the recent advances of OAM generation in fiber systems are reviewed, which are expected to make a contribution to space-division multiplexing optical fiber transmission systems, atom manipulation, microscopy, and so on.

Abstract: Orbital angular momentum (OAM) beams, characterized by the helical phase wavefront, have received significant interest in various areas of study. There are many methods to generate OAM beams, which can be roughly divided into two types: spatial methods and fiber methods. As a natural shaper of OAM beams, the fibers exhibit unique merits, namely, miniaturization and a low insertion loss. In this paper, we review the recent advances in fiber OAM mode generation systems, in both the interior and exterior of the beams. We introduce the basic concepts of fiber modes and the generation and detection theories of OAM modes. In addition, fiber systems based on different nuclear devices are introduced, including the long-period fiber grating, the mode-selective coupler, microstructural optical fiber, and the photonic lantern. Finally, the key challenges and prospects for fiber OAM mode systems are discussed.

Keywords: orbital angular momentum; long period fiber grating; mode selective coupler; photonics lantern; microstructure optical fiber

---

1. Introduction

Since Allen first demonstrated the orbital angular momentum (OAM) of light as an independent dimension in 1992 [1], beams carrying OAM have attracted increasing interest in various fields. OAM beams are characterized by the phase singularity and helical wavefront. Due to the helical wavefront, the propagating direction (or wavevector) is variant with the azimuthal angle, which is also helical relative to the optical axis. The helical degree is described by so-called "topological charge (TC)". OAM beams with different TC can be regarded as several independent dimensions carrying information. These properties make OAM beams different from conventional plane light waves and many unique applications in terms of atom manipulation [2–4], nanoscale microscopy [5], optical tweezers [6–8], optical communication [9–13], and data storage [14,15] have been realized.

There are many methods used to generate OAM beams. Such methods can be roughly divided into two categories, spatial and fiber generating methods. Spatial methods are generally assisted by spatial light modulators [16], spiral phase plates [17,18], diffractive phase holograms [19–22], metamaterials [23–26], cylindrical lens pairs [27], q-plates [28,29], photonics integrated circuits including micro-ring resonators [30], among other devices, as shown in Figure 1. Each spatial method has its own advantages and disadvantages. However, there are two common defections for them, i.e., the high insertion loss and the large volume. Because there exists interface with high refractive index differences in these methods, the conversion processes from incident beams to OAM beams are not modest, thereby leading to a relatively high insertion loss. Meanwhile, the volumes of spatial devices are usually large. This indicates that miniaturization and integration of the space systems is challenging.

Compared with spatial generating methods, fiber generating methods have shown their certain advantages. A converter applied in fiber methods can be fiber gratings [31–44], mode selective couplers [45–48], and microstructure optical fibers [49–51] and photonic lanterns [52,53]. As a cylindrical waveguide, fiber is a natural beam shaper for OAM beams. Incident beams of any shapes will be converted into the eigenmodes with cylindrical symmetry under the restriction of fiber. Since the conversion process is more modest than that in spatial generating methods, the energy efficiency is higher in fiber generating methods. Moreover, all the devices in fiber generating methods are smaller, which greatly facilitates miniaturization.

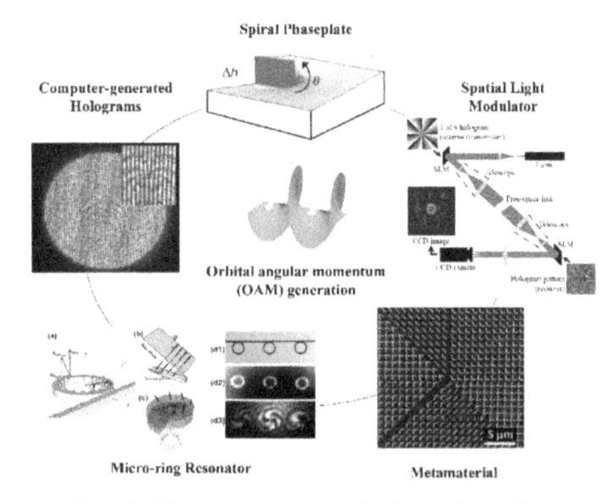

Figure 1. A summary of typical OAM generation techniques based on spatial components (spiral phase plate [18], spatial light modulator [16], metamaterial [24], micro-ring resonator [30], and computer-generated holograms [22]) [16–30]. Reprinted with permission from [18], copyright 2010 Springer Nature; [16], copyright 2004 The Optical Society; [24], copyright 2011 AAAS; [30], copyright 2012 AAAS; [22], copyright 2011AAAS.

This paper highlights recent advances in fiber OAM generation systems. We start by introducing three types of fiber modes in Section 2, i.e., cylindrical vector (CV), linearly polarized (LP), and OAM modes. Following this, Section 3 briefly describes the basic concepts and theories of OAM beam generation and detection. In Section 4, the recent advances in fiber OAM generation systems are given. A brief discussion of this research and further expectations are presented in Section 5.

2. Three Types of Fiber Mode

2.1. Cylindrical Vector Modes

Cylindrical vector (CV) modes, whose polarization varies based on the spatial location, show many unique properties and applications compared with the conditional plane light waves. As the eigenmodes in fiber, any mode field in fiber can be regarded as the superposition of CV modes, with different amplitudes and phase differences. CV modes are divided into different azimuthal orders. Each azimuthal order mode is composed of two or four degenerated modes, whose propagation constants are almost the same. For general step index fibers, the zeroth azimuthal order mode, which is only composed of two degenerated modes, namely, $HE_{1,m}^{even}$ and $HE_{1,m}^{odd}$ mode. The first order mode is composed of $TM_{0,m}$, $TE_{0,m}$, $HE_{2,m}^{even}$, and $HE_{2,m}^{odd}$ mode. The lth ($l > 1$) order mode is composed of $EH_{l-1,m}^{even}$, $EH_{l-1,m}^{odd}$, $HE_{l+1,m}^{even}$, and $HE_{l+1,m}^{odd}$. Here, m is the radial order, denoting the number of noaxial radial nodes of the mode. In general, researches are simply conducted on the properties of radial order $m = 1$. We use the default $m = 1$ in this paper unless indicated. Figure 2a,b give the propagating properties and time average intensity pattern of two typical CV modes, TE_{01} and TM_{01}. As shown in Figure 2a,b, the polarization state of each point is linearly polarized, and the polarized direction of each point is related to the spatial angle. In the center of CV mode, there is a so-called "polarized singularity", where the intensity vanishes in the area. This is because the radial distribution is defined by the cylindrical function with higher azimuthal order ($l > 0$), such as the Bessel function, Laguerre Gaussian (LG) function, and Hermite-Gaussian (HG) with non-zeroth orders. These functions with non-zeroth orders are zero at the center point. The last column are the time average patterns during an integer-number period. Because the response frequency of a detective device is much slower than the frequency of light, what we detect using the detective devices are the time average intensity patterns during countless periods of light, which is close to the integer-period time average patterns. The integer-period time average intensity patterns of CV modes are doughnut shaped for a higher azimuthal order mode (and biscuit shaped for zeroth order mode).

2.2. Linearly Polarized Modes

Linearly polarized (LP) modes are another group of fiber mode base. Their polarization states are also linearly polarized, but the polarized directions are the same, invariant with the spatial angle. However, the polarized amplitude at each point changes periodically with the spatial angle. If considering the radial field distribution simultaneously, the intensity patterns of LP modes are $2l$ lobes, where l is the azimuthal order, as mentioned for CV modes. LP modes are not the eigenmodes in fiber unless the four degenerated modes are strictly degenerated. However, on the end facet of fiber, the field can still be expressed in LP mode bases because there is no restriction of cylindrical waveguide outside the fiber. Any lth order electric field in fiber can also be decomposed into four LP mode bases with different complex amplitudes. Figure 2c,d show the propagating properties and time average intensity patterns of two typical LP modes LP_{11}^{even} and LP_{11}^{odd}, where $\hat{x}$ denotes the linear polarization along the x-axis and $LP_{l,m}^{even}$ and $LP_{l,m}^{odd}$ represent the two LP modes with complementary intensity patterns. Here $\hat{x}$ can be replaced by other linearly polarized symbols when the observed coordinates change.

2.3. Orbital Angular Momentum Modes

OAM modes, are also one of the groups of fiber-based modes. Unlike the conventional plane light wave, OAM modes are characterized by a helical phase front $e^{\pm il\xi}$ [1], where $\pm\, l$ is the TC and ξ is the azimuthal angle related to the optic axis. In addition, l can take the integer numbers from zero to $+\infty$. It should be noted that l is the same as the azimuthal order of CV modes. For different points on the beam cross-section with the same radius, the polarization states are the same, but with different phases. This indicates the helical phase front of OAM modes. Figure 2e,f show the propagating properties and time average intensity patterns of two typical OAM modes $\hat{\sigma}^{-}OAM_{+1}$ and $\hat{x}OAM_{+1}$. Taking $\hat{\sigma}^{-}OAM_{+1}$ as an example, as indicated in Figure 2e, electric vectors at each point with the same radius on the beam cross section are the right-hand circular polarized ($\hat{\sigma}^{-}$). The phase factor of OAM_{+l} should be $e^{i(kz-\omega t+l\xi)}$. For lth order OAM modes, the number of equal phase points on the beam cross section will be l. As shown in Figure 2e, initially, the x-polarization point is located at $\xi = 0$ $(kz - \omega t + \xi = 0)$. Then, when the field propagates to $kz - \omega t = \frac{\pi}{4}$, the x-polarization point (with the same phase) is located at $\xi = -\frac{\pi}{4}$ $(kz - \omega t + \xi = 0)$. This means that, with $\xi = -(kz - \omega t)$, the equal phase point appears along the clockwise direction during the propagating, which indicates the factor $e^{i\xi}$. Thus, Figure 2e indicates $\hat{\sigma}^{-}OAM_{+1}$ mode. The analysis method is similar to that of $\hat{x}OAM_{+l}$. The symbol $\hat{x}$ simply indicates the linear polarization, which can be substituted by another linear polarization symbol when the observation coordinates rotate.

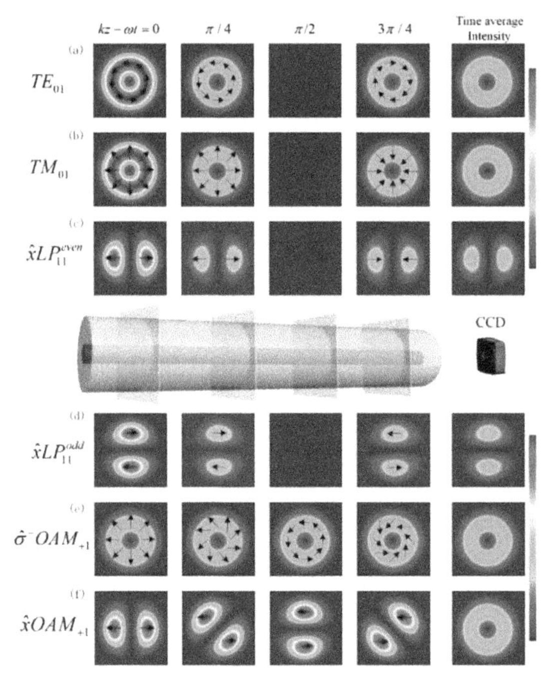

Figure 2. The propagation properties during half a period and the integer-period time average patterns of two typical CV modes, (**a**) TE_{01} and (**b**) TM_{01}; and two typical LP modes, (**c**) $\hat{x}LP_{11}^{even}$ and (**d**) $\hat{x}LP_{11}^{odd}$; and two typical OAM modes, (**e**) $\hat{\sigma}^{-}OAM_{+1}$ and (**f**) $\hat{x}OAM_{+1}$.

For a typical combination of CV modes used to generate the OAM mode, $\hat{\sigma}^{-}OAM_{+1} = TM_{01} - iTE_{01}$, as shown in Figure 2 The term "$-iTE_{01}$" indicates the figures of the first row (Figure 2a) with a $-\frac{\pi}{2}$ phase delay. The physical meaning of $\hat{\sigma}^{-}OAM_{+1} = TM_{01} - iTE_{01}$ is the interference between TM_{01} and TE_{01} patterns with a $-\frac{\pi}{2}$ phase delay of TE_{01}. In Figure 2, when TM_{01} propagates to $kz - \omega t = \frac{\pi}{2}$, TE_{01} reaches $kz - \omega t = 0$. Adding these two electric fields, we obtain the $\hat{\sigma}^{-}OAM_{+1}$ at $kz - \omega t = \frac{\pi}{2}$. In addition to $\hat{\sigma}^{-}OAM_{+1} = TM_{01} - iTE_{01}$, there are a series of transformation relations among CV, LP, and OAM modes. Moreover, we may note that there is no difference among the time average intensity patterns of CV and OAM modes. To further assure the phase information, a fundamental mode is usually used to interfere with a higher order fiber mode. Through the interference patterns we can obtain the phase information to confirm the specific electric vector field of the same doughnut intensity patterns. In the following section, we are going to derive the entire relation for CV modes, LP modes, and OAM modes, and introduce the generation and detection of OAM modes.

3. Basic Concepts and Theories of OAM Beams Generation and Detection

3.1. Transformation Relation among CV Modes, LP Modes, and OAM Modes

As the three groups of fiber mode bases, there is a transformation relation among CV modes, LP modes and OAM modes. Among these three modes, CV modes are the eigenmodes in fiber, which are able to propagate stably in fiber. We start our introduction of the transformation relation with CV modes.

In an axisymmetric index profile fiber, the intrinsic electric field is under the restriction of a cylindrical waveguide. By solving the Helmholtz equation in cylindrical coordinates, the eigenmodes in fiber can be derived, that is the CV modes. The solutions are given in Equation (1):

$$\begin{pmatrix} E_x(r,\xi,z) \\ E_y(r,\xi,z) \end{pmatrix} = \begin{cases} F_{l,m}(r)\begin{pmatrix} \cos(l\xi) \\ \sin(l\xi) \end{pmatrix} e^{i\beta_1 z}; & EH^{even}_{l-1,m}/TM_{0,m} \\ F_{l,m}(r)\begin{pmatrix} -\sin(l\xi) \\ \cos(l\xi) \end{pmatrix} e^{i\beta_2 z}; & EH^{odd}_{l-1,m}/TE_{0,m} \\ F_{l,m}(r)\begin{pmatrix} \cos(l\xi) \\ -\sin(l\xi) \end{pmatrix} e^{i\beta_3 z}; & HE^{even}_{l+1,m} \\ F_{l,m}(r)\begin{pmatrix} \sin(l\xi) \\ \cos(l\xi) \end{pmatrix} e^{i\beta_4 z}; & HE^{odd}_{l+1,m} \end{cases} \tag{1}$$

where ξ is the spatial angle, $F_{l,m}(r)$ is radial field distribution, l is the azimuthal order of CV modes, m is the radial order and $\beta_{1\text{-}4}$ are the propagation constants. Usually, $F_{l,m}(r)$ is the Bessel function in a step index fiber. For l = 1, $EH^{odd}_{l-1,m}$ should be substituted by $TM_{0,m}$ and $EH^{even}_{l-1,m}$ should be substituted by $TE_{0,m}$. In this section, however, we simply use $EH^{even}_{l-1,m}$ and $EH^{odd}_{l-1,m}$ to express the corresponding CV modes for conciseness, even for l = 1. Any lth-order electric field in fiber can be decomposed into the superposition of the four degenerated eigenmodes, that is, $E = AEH^{even}_{l-1,m} + BEH^{odd}_{l-1,m} + CHE^{even}_{l+1,m} + DHE^{odd}_{l+1,m}$, where $(A\,,B\,,C\,,D)^T$ is an arbitrary complex vector. The amplitudes and the phases of $(A\,,B\,,C\,,D)^T$ represent the amplitudes and the relative phases of $EH^{even}_{l-1,m}$, $EH^{odd}_{l-1,m}$, and $HE^{even}_{l+1,m}$, $HE^{odd}_{l+1,m}$, respectively.

In OAM mode bases, the four mode bases are $\hat{x}OAM_{-l}, \hat{y}OAM_{-l}, \hat{x}OAM_{+l}$, and $\hat{y}OAM_{+l}$, where $\hat{x}(\hat{y})$ represents the $x(y)$ linearly polarized direction and $OAM_{\pm l}$ represents the OAM modes with TCs $\pm l$. With the aid of Jones calculus, any lth-order electric field can be expressed as $E = OAM_{-l}\begin{pmatrix} x_{-l} \\ y_{-l} \end{pmatrix} + OAM_{+l}\begin{pmatrix} x_{+l} \\ y_{+l} \end{pmatrix}$. Likely, as an arbitrary complex vector, $(x_{-l}, y_{-l}, x_{+l}, y_{+l})^T$ completely describes the entire lth-order electric field. In LP mode bases, the four mode bases are $\hat{x}LP^{even}_{l,m}, \hat{y}LP^{even}_{l,m}, \hat{x}LP^{odd}_{l,m}$, and $\hat{y}LP^{odd}_{l,m}$, where $LP^{even}_{l,m}$ and $LP^{odd}_{l,m}$ represent the two LP modes with complementary intensity patterns. Any lth-order electric field can be expressed in LP mode bases as $E = LP^{even}_{l,m}\begin{pmatrix} x_e \\ y_e \end{pmatrix} + LP^{odd}_{l,m}\begin{pmatrix} x_o \\ y_o \end{pmatrix}$. The corresponding complex vector is $(x_e, y_e, x_o, y_o)^T$.

For three groups of fiber mode bases, there exists transformation relation from CV modes to LP modes and OAM modes as follows (the detailed derivation can be found in [54]):

$$\frac{1}{2}\begin{pmatrix} 1 & -i & 1 & i \\ i & 1 & -i & 1 \\ 1 & i & 1 & -i \\ -i & 1 & i & 1 \end{pmatrix}\begin{pmatrix} A \\ B \\ C \\ D \end{pmatrix} = \begin{pmatrix} x_{-l} \\ y_{-l} \\ x_{+l} \\ y_{+l} \end{pmatrix} \tag{2}$$

for CV modes to OAM modes:

$$\begin{pmatrix} 1 & 0 & 1 & 0 \\ 0 & 1 & 0 & 1 \\ 0 & -1 & 0 & 1 \\ 1 & 0 & -1 & 0 \end{pmatrix}\begin{pmatrix} A \\ B \\ C \\ D \end{pmatrix} = \begin{pmatrix} x_e \\ y_e \\ x_o \\ y_o \end{pmatrix} \tag{3}$$

for CV modes to LP modes, and:

$$\frac{1}{2}\begin{pmatrix} 1 & 0 & i & 0 \\ 0 & 1 & 0 & i \\ 1 & 0 & -i & 0 \\ 0 & 1 & 0 & -i \end{pmatrix}\begin{pmatrix} x_e \\ y_e \\ x_o \\ y_o \end{pmatrix} = \begin{pmatrix} x_- \\ y_- \\ x_+ \\ y_+ \end{pmatrix} \tag{4}$$

for LP modes to OAM modes. If we define the transformation matrix of CV modes to LP modes (Equation (3)) as M_1, and the transformation matrix of LP modes to OAM modes as M_2 (Equation (4)), then the transformation matrix of CV modes to OAM modes (Equation (2)) is just the matrix product of M_1 and M_2, namely, $M_3 = M_1M_2$. Figure 3 shows the intuitive sketch describing the transformation relation among CV modes, LP modes and OAM modes. In Figure 3, the circle indicates all lth order electric fields, which should be a four-dimensional complex space, however, we simply use a two-dimensional circle to express them. The lines with different colors in the three circles indicate the different divided methods of the four-dimensional complex space. Notice that, the shapes of the lines are irrelevant. They are simply used to clearly denote the different space divided methods. Four degenerated modes operate as the different bases to completely describe the four-dimensional complex space (the entire lth order modes), where they can be transformed into each other through the transformation matrix derived above. The transformation among the mode bases is equivalent to the base transformation in the four-dimensional complex space, by the aforementioned matrices. The generation of pure OAM modes, is equivalent to adjust the mode field using physical methods to further simplify the expression in OAM mode bases. For example, a pure OAM mode $\hat{x}OAM_{-l}$ expressed in OAM mode bases is $(1,0,0,0)^T$, whereas $0.5\left(EH^{even}_{l-1,m} + iEH^{odd}_{l-1,m} + HE^{even}_{l-1,m} - iHE^{odd}_{l-1,m}\right)$ or $0.5(1,i,1,-i)^T$ in CV mode bases and $\hat{x}LP^{even}_{l,m} - i\hat{x}LP^{odd}_{l,m}$ or $(1,-i,0,0)^T$ in LP mode bases. The three expressions denote the same spatial field which is not a pure CV mode or pure LP mode because the expression in the corresponding mode bases contain several components. The generation of pure CV modes or LP modes is similar.

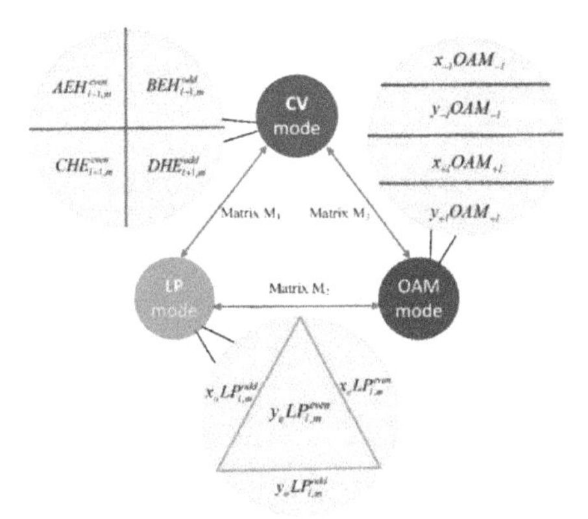

Figure 3. Sketch of the four-dimension complex space model and the transformation among CV mode, LP mode, and OAM mode bases of lth azimuthal order modes.

3.2. Generation of OAM Beams

Recently, all-fiber OAM generation methods have received increasing interests owing to the advantages of the lower insertion loss and better compatibility with the optical fiber communication links. The all-fiber system used to generate OAM beams can be summarized as shown in Figure 4. The system is divided into three parts: mode couple module, field control module, and polarization separation module.

Due to the difference of the effective refractive index (ERI) between different order modes being sufficiently large that mode coupling between different mode groups cannot occur by perturbations, such as fiber bending, twisting or extrusion. The mode couple module is used to couple the fundamental mode to a specific lth-order CV mode. It is usually composed of the fiber gratings [31–44] and the fiber couplers [45–48]. As mentioned above, the lth-order CV modes are composed of four degenerated modes. The mode couple module couples the fundamental mode to the lth-order CV modes, which can be seen as the superposition of the initial four degenerated modes with random amplitudes and phases in reality. This typically can't be used to generate pure OAM beams.

The field control module, the polarization controller (PC) usually used in the fiber system, is applied to redistribute the generated random state of lth-order CV modes. Due to the ERIs of the four degenerated modes with the same order are approximately the same, the four degenerated modes with the same order will be strongly coupled to other degenerated modes in the same mode group when passing through the PC because of the bending, twisting, and extruding of the few-mode fiber (FMF) provided by the PC. This will change the relative amplitudes and phases among these four degenerated modes. In some particular situation, the pure OAM modes can be generated, such as $\hat{\sigma}^{\pm}OAM_{\pm l} = HE^{even}_{l+1,m} \pm iHE^{odd}_{l+1,m}$.

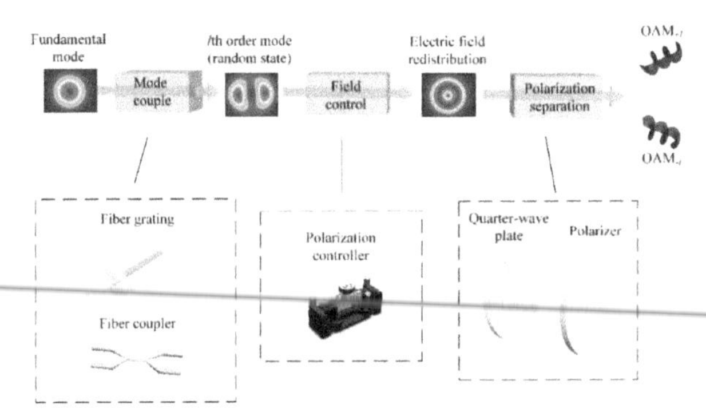

Figure 4. Sketch of fiber OAM modes generation system. Reprinted with permission from [54], copyright 2018 De Gruyter.

The polarization separation module is used to separate two orthogonal polarized OAM modes after the polarization control module. It is composed of a QWP and a polarizer with a particular angle. The angle depends on the mode distribution prior to the polarization separation module. If the field control module generates an electric field of two orthogonal polarized OAM modes which carry different TCs, such as $\hat{\sigma}^{+}OAM_{-l} + \hat{\sigma}^{-}OAM_{+l}$, the two polarized orthogonal OAM modes can be separated through a polarization separation module. A polarization separation module is not necessary if the electric field after the polarization control module is carrying pure TC.

3.3. Detection of OAM Beams

Beams from the fiber will be converted into a divergent wave once they leave the fiber. For conciseness, we use a spherical wave to discuss the same physical process in this review. Indeed, they could be other types of divergent waves besides spherical waves, as long as their wavevectors are different from the other interference beam, such as a Gaussian light, whose curvatures of these lights are not the same on the beam cross-section when propagating while a spherical wave is a light wave with an invariant curvature. Without loss of generality, the physical meaning of the discussion below will not change under the assumption of a spherical wave.

The spherical wave is characterized by the factor e^{ikR}, where k is the wavevector and $R = (x^2 + y^2 + z^2)^{\frac{1}{2}}$ is the radius relative to the light source. A spherical wave carrying OAM TC = +

l can be expressed as $E_s(r,\xi,z) = F_{l,m}(r)e^{i(l\xi+k_s(r^2+z^2)^{\frac{1}{2}})}$, where k_s denotes the wavevector of OAM mode, $r = (x^2+y^2)^{\frac{1}{2}}$ and ξ denote the transverse radius and the spatial angle on the beam cross section respectively and z denotes the propagating distance along with the optics axis. In addition, $F_{l,m}(r)$ is the radial field distribution, as mentioned above. A fundamental mode with a different divergence is usually used to interfere with OAM beam to detect the phase information of OAM beams. A spherical fundamental mode can be expressed as $E_f(r,\xi) = F_{0,1}(r)e^{ik_f(r^2+z^2)^{\frac{1}{2}}}$, where k_f denotes the wavevector of fundamental mode. In general, the radial field distribution only affects the spot size but do not contribute to the final interference shapes. Herein, we omit the discussion about the radial field. Assume that the polarizations of the signal OAM beams and the interference beams are the same, the interference of the electric field should be:

$$|E|^2 = \left(E_s + E_f\right)^*\left(E_s + E_f\right) = |E_s|^2 + \left|E_f\right|^2 + E_s^*E_f + E_f^*E_s \tag{5}$$

where the left term $|E_s|^2 + \left|E_f\right|^2$ is the direct current (DC) component invariant with the spatial angle ξ, which acts as an intensity base in the interference patterns. Because the DC component doesn't affect the shapes of the patterns, we emphatically show the interference term $E_s^*E_f + E_f^*E_s$ through Equation (6):

$$E_s^*E_f + E_f^*E_s \propto F_{l,m}(r)F_{0,1}(r)\cos\left(l\xi + \left(k_s - k_f\right)\left(r^2+z^2\right)^{\frac{1}{2}} + \phi_s - \phi_f\right) \tag{6}$$

where the term $F_{l,m}(r)F_{0,1}(r)$ indicates the area of interference related to the spot sizes of OAM beam and fundamental beam. Here, $\phi_s - \phi_f$ indicates the initial phase difference of OAM beam and the fundamental beam. Based on Equation (6), Figure 4 shows the interference patterns of OAM_{+1}, OAM_{+2} and OAM_{+3} with the different wavevector difference $k_s - k_f$ and initial phase difference $\phi_s - \phi_f$.

As can be seen, when $k_s \neq k_r$, the patterns exhibit the vortex shapes, while the vortex number indicates $|l|$, the absolute value of TC. Meanwhile, the vortex rotation direction is closely related to the difference between wave vectors of two beams. When $k_s - k_f < 0$, which indicates that the fundamental beam is more divergent, a counter-clockwise vortex indicates l as the positive value while a clockwise vortex indicates the negative value. Contrary patterns exist when $k_s - k_f > 0$. When $k_s = k_f$, the interference patterns do not exhibit the vortex shapes and we are unable to judge the specific TC for this type of interference.

Moreover, Figure 5a,b show the effects of the initial phase difference $\phi_s - \phi_f$ between the two interference beams on the interference patterns. When the initial phase difference changes, the patterns simply rotate an angle but do not change the number and direction of the vortex. This means that the initial phase difference does not disturb OAM beam detection under the interference condition.

If the physical process is angle-independent, the distinction between OAM_{+l} and OAM_{-l} is not important. We can simply define one direction of the rotated vortex as OAM_{+l} and the other as OAM_{-l}. The counter-clockwise rotated vortex is usually defined as OAM_{+l}. However, if the physical process is related to the meta-surface, chiral devices, and some other angle-dependent devices, the distinction between OAM_{+l} and OAM_{-l} is crucial, and a meticulous method must be used, as discussed above. That is, the divergence of the signal OAM beam and the interference fundamental beam must be given.

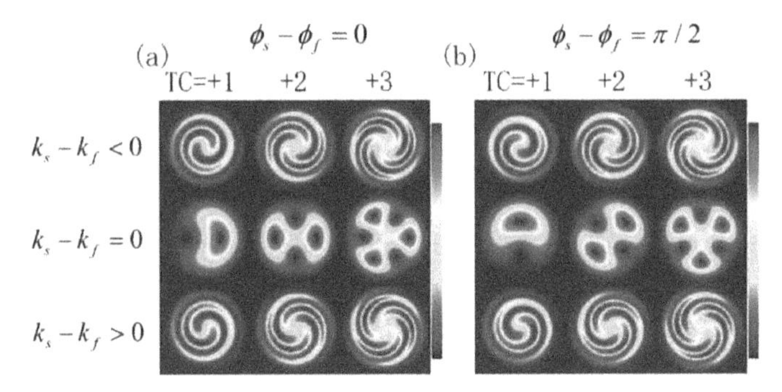

Figure 5. The interference patterns of OAM_{+1}, OAM_{+2} and OAM_{+3} when $k_s - k_f > 0$, $k_s - k_f = 0$ and $k_s - k_f < 0$ and (**a**) $\phi_s - \phi_f = 0$ and (**b**) $\phi_s - \phi_f = 0$.

Figure 6a shows another influencing factor in addition to the difference in wavevector, that is, the tilted degree, which indicates the difference in the propagating direction of the reference beam from the signal OAM beam. The larger the tilted degree is, the larger the angle of the propagating direction between the OAM beam and the reference beam. The effects of the TC and wavevector have been discussed above (Figure 5). Here, we highlight the effect of the tilted degrees. In Figure 6a, we set the propagating direction of the signal OAM beam as the reference axis (red arrow). On the top of Figure 6a, the red arrow indicates that the propagation direction of the signal OAM beam and the reference beam are the same. In addition, black arrows indicate the tilted degree between these two beams. As can be seen, the interference patterns shift into a fork-like shape from the vortex shape when the tilted degree increases. In addition, the open direction of the "fork" is closely related to the tilted direction. As shown in the second row in Figure 6a, for OAM_{-1}, the fork is upward open when the fundamental mode is left-tilted in relation to signal OAM beam. While it is downward open direction for a right-tilted fundamental mode. For OAM_{+1} (the first row of Figure 6a), the results are opposite, compared to the first and the second rows in Figure 6a. Thus, interference patterns are closely related to the left-tilted or right-tilted between the OAM mode and the fundamental mode.

As for the effect of the difference of wavevectors, compared with the second and the fourth row of Figure 6a, the divergence (wavevector) difference $k_s - k_f$ does not disturb the judgement for nonzero tilted degree. Unlike a case of normal incidence (the middle red arrow), the vortex shape is different when the divergence difference $k_s - k_f$ changes, as shown in the middle column in Figure 6a, which will disturb the judgment about the TC of signal OAM beam. In addition, when disregarding TC = $+l$ or TC = $-l$ of the OAM mode, upward-open-fork shape patterns are usually defined as $+l$ while the downward-open-fork shape patterns are defined as $-l$. Otherwise, it's necessary to introduce the reference beam is left-tilted or right-tilted relative to the propagation direction of signal OAM beam.

In addition to using a fundamental beam with a different divergence to interfere with OAM beams, there are still some other methods to obtain the TC information of an OAM beam. The reference light E_f can be other types, such as a tilted plane light wave and an OAM beam itself with a spatial translation. Without proof, we provide the typical interference patterns of these two types.

Figure 6b shows the interference between a non-divergent OAM beam (OAM_{-1}, OAM_{-2} and OAM_{-3}) and right-tilted plane light wave ($k_s = k_f = 0$). This is the degenerated case when $k_s - k_f = 0$ and titled degree is non-zero in Figure 6a. We are able to judge different TCs through this interference method. In physical angle-independent processes, people generally define the number of interference line in the upper part l more than that in the lower part. In other words, a right-tilted reference beam is usually assumed if not caring about the sign of the TC. For an angle-dependent physical process, the tilted degree between the OAM mode and fundamental mode must be given.

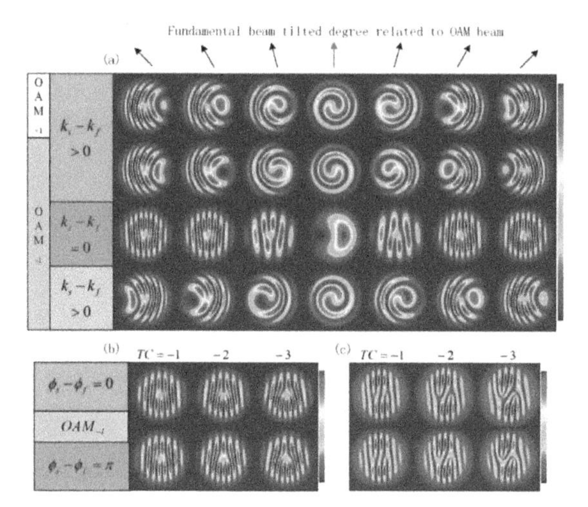

Figure 6. (**a**) Interference patterns with different TCs, wave vectors and tilted degrees. The interference patterns of OAM beams ($OAM_{-1}, OAM_{-2}, OAM_{-3}$) with (**b**) the tilted plane light wave; (**c**) itself with a translation when the initial phase interference $\phi_s - \phi_f = 0$ or π.

As for Figure 6c, the OAM beam interferes with itself with a translation. If setting the propagating direction of the left-translational OAM beam as the reference axis (Figure 6a), the other OAM beam operates as a left-tilted reference beam related to the signal OAM beam. These two interference beams have the same divergence (wavevector). Thus, in the left area of arbitrary subfigure in Figure 6c, the TC is recognized as +*l* for a downward open fork-shaped pattern and −*l* for upward open fork-shaped pattern, according to the judging method given in Figure 6a. For the right area, the result is opposite, where the TC is +*l* for upward-opening-fork shape pattern and −*l* for downward-opening-fork shape pattern. To integrate these two methods, when comparing Figure 6b,c, we just need to focus on the right part in Figure 6c and use the same judgment process as that used for the tilted plane light wave (Figure 6b). In addition, the two rows patterns in Figure 6b,c show the effect of initial phase difference on the interference pattern. This simply leads to a translation of the patterns but not disturb the judgment about the TC.

4. Advances in Fiber OAM Generation Systems

4.1. Fiber Grating-Based OAM Generation Systems

Optical fiber gratings, a diffraction grating formed by the axially periodic refractive index modulation of fiber core using a certain method, have been developed into a mature technology and have been widely used in optical communications and fiber sensing in recent decades [55]. It is well-known that optical fiber gratings can be classified into two types, that is, fiber Bragg grating (FBG) [56] and long-period fiber grating (LPFG) [57–68], according to the grating period. The grating period of FBG is usually hundreds of nanometers or a few micrometers and the grating period of LPFG is usually hundreds of micrometers. In the FBG or LPFG, the fundamental mode in fiber core is coupled to the backward- or forward-propagating cladding mode or core mode at distinct wavelengths respectively. LPFG, written into a few-mode fiber, is the most commonly used an OAM conversion device. The working principle of the LPFG is the coupled mode theory. When the grating period satisfies the phase-matching condition $\lambda_m = (n_{eff1} - n_{eff2})\Lambda$, where λ_m is the resonant wavelength, n_{eff1} and n_{eff2} are the effective refractive index (ERI) of fundamental mode and higher-order core mode, respectively, the fundamental mode in the fiber core will be coupled to higher-order core mode, thereby leading to the transmission spectrum of the fundamental mode with one or a few notches, as shown in Figure 7c.

There have been several methods used to fabricate an LPFG, including ultraviolet laser [57], femtosecond laser irradiation [58], CO_2 laser irradiation [27,32,43,59–61], cleaving-splicing method [62], arc discharge [63], acousto-optic interaction [64], mechanical micro-bending [31,33–35,65], and so on. The CO_2 laser irradiation and mechanical micro-bending are the two frequently-used methods to fabricate the LPFG to generate the OAM modes.

The working principle of the mechanically induced LPFG is shown in Figure 7a. The few-mode fiber is placed between two plates with periodically grooves or between a plate with periodically grooves and a flat plate. When the period satisfies the phase-matching condition and putting pressure on the one of the two plates, the fundamental mode will be coupled to the higher-order core mode. The advantages of this method are that one can adjust the resonant wavelength or the order of coupled higher order mode, by adjusting the grating period. In addition, the length of the fiber under pressure, which controls the linewidths of the notch, can easily be changed. In addition, the depth of the notches can be tuned by adjusting the pressure. Finally, when the pressure is removed, the transmission spectrum of the fiber returns to its initial spectrum. The disadvantage of this method is some fiber cannot be used to fabricate an LPFG.

Compared with the methods mentioned above, a high-frequency CO_2 pulse is an attractive tool used to fabricate an LPFG owing to its convenience, economy and high efficiency. As shown in Figure 7b, the CO_2 laser irradiation method uses a high-frequency CO_2 laser to continuously notch grooves on one side of the few-mode fiber to form a series of periodic structures. Because of the advantages of a CO_2 laser writing technique, the LPFGs can be successfully written into various types of special fiber, such as multi-core fiber [66] and photonic crystal fiber [61,67], and different types of LPFG, such as a sampled or a chirped grating, can be inscribed in the fiber [68]. However, this method has broken the fiber structure, and the resonant wavelength and the coupled mode are fixed once the LPFG is fabricated.

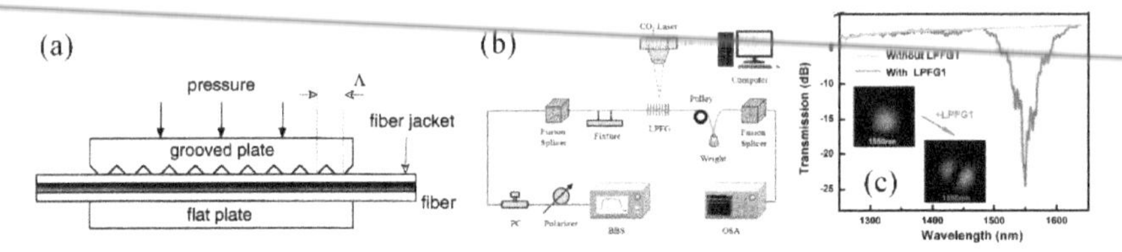

Figure 7. Experimental setup of (**a**) mechanical micro-bending and (**b**) CO2 laser irradiation method to fabricate LPFGs; (**c**) transmission spectrum of the LPFGs. Reprinted with permission from [65], copyright 2000 The Optical Society; [60], copyright 2018 The Optical Society; [43], copyright 2018.

In general, fiber grating-based methods for pure OAM mode generation, with experimental set up as shown in Figure 8a, can be classified into two types according to the source modes. For one type, the OAM modes can be generated by combining two linearly polarized modes with a $\pi/2$ phase shift, where the generated OAM modes have no spin angular momentum (SAM). For the other type, the OAM modes can be generated by combining two vector modes ($HE_{l+1,m}$ or $EH_{l-1,m}$) with a $\pi/2$ phase shift, where the generated OAM modes have the SAM at the same time. There have been many reports about the generation of these two types OAM modes using LPFGs [32,34,35,37,43].

For the first type, namely the generation of linearly-polarized OAM (LP-OAM) modes, the LP-OAM modes in the x and y polarized directions can be obtained by eliminating other component in the OAM mode bases except one, such as $(x_{-l}, y_{-l}, x_{+l}, y_{+l})^T = (1,0,0,0)^T$, which corresponds to $\hat{x}OAM_{-l}$, an x-linear polarized OAM beam with TC = +*l*. The expression in other mode bases can be obtained by substituting this vector into Equation (2) or Equation (4). $(x_{-l}, y_{-l}, x_{+l}, y_{+l})^T = (1,0,0,0)^T$ is equivalent as $(A,B,C,D)^T = 0.5, 0.5i, 0.5, -0.5i^T$ and $(x_e, y_e, x_o, y_o)^T = (1,0,-i,0)^T$. Given a physical meaning, that is $\hat{x}OAM_{-l} = 0.5EH^{even}_{l-1,m} + 0.5iEH^{odd}_{l-1,m} + 0.5HE^{even}_{l+1,m} - 0.5iHE^{odd}_{l+1,m} = \hat{x}LP^{even}_{l,m} - i\hat{x}LP^{odd}_{l,m}$. Thus, the LP-OAM can be obtained by combining the even and odd $LP_{l,1}$ mode having the same polarized direction with a $\pm\pi/2$ phase shift. The first order LP-OAM modes, composed of LP_{11}^{a} and LP_{11}^{b} mode with a $\pi/2$ phase shift, using a two-mode fiber (TMF), a mechanical LPFG, a mechanical rotator and metal slabs, as shown in Figure 8a, and the second order LP-OAM modes, composed of LP_{21}^{a} and LP_{21}^{b} mode with a $\pi/2$ phase shift, utilizing an LPFG written in a four-mode fiber induced by CO_2 laser irradiation, have been experimentally demonstrated in [35] and [37], respectively.

For the other type, namely the generation of circularly-polarized OAM (CP-OAM) modes, the typical CP-OAM modes with SAM being ±1 can be obtained by the following equation [43]:

$$\begin{pmatrix} \hat{\sigma}^{-}OAM_{+1} \\ \hat{\sigma}^{+}OAM_{-1} \\ \hat{\sigma}^{+}OAM_{+1} \\ \hat{\sigma}^{-}OAM_{-1} \end{pmatrix} = F_{1,1}(r)\begin{pmatrix} 1 & i & 0 & 0 \\ 1 & -i & 0 & 0 \\ 0 & 0 & 1 & i \\ 0 & 0 & 1 & -i \end{pmatrix}\begin{pmatrix} HE_{21}^{even} \\ HE_{21}^{odd} \\ TM_{01} \\ TE_{01} \end{pmatrix}, l=1$$
$$\begin{pmatrix} \hat{\sigma}^{-}OAM_{+l} \\ \hat{\sigma}^{+}OAM_{-l} \\ \hat{\sigma}^{+}OAM_{+l} \\ \hat{\sigma}^{-}OAM_{-l} \end{pmatrix} = F_{l,1}(r)\begin{pmatrix} 1 & i & 0 & 0 \\ 1 & -i & 0 & 0 \\ 0 & 0 & 1 & i \\ 0 & 0 & 1 & -i \end{pmatrix}\begin{pmatrix} HE_{l+1,1}^{even} \\ HE_{l+1,1}^{odd} \\ EH_{l-1,1}^{even} \\ EH_{l-1,1}^{odd} \end{pmatrix}, l\geq 2 \tag{7}$$

where $\hat{\sigma}^{+}$ and $\hat{\sigma}^{-}$ represent left- and right-handed circular polarization. Taking $\hat{\sigma}^{+}OAM_{+1}$ as an example, one can just set the expression in the OAM mode bases as $(x_{-l}, y_{-l}, x_{+l}, y_{+l})^{T} = (0,0,1,i)^{T}$ and calculate the corresponding expression in the CV mode bases or LP mode bases, the results of which are the same as Equation (7). The first- and second-order CP-OAM modes composed of even and odd HE_{21}, HE_{31}, EH_{11} and TM_{01}, TE_{01} modes with a $\pi/2$ phase shift are experimentally demonstrated in [43] by our group, as shown in Figure 8b,c, using two LPFGs written in a four-mode fiber induced by CO_2 laser. The first LPFG is used to convert fundamental mode to the first order mode. The second LPFG is used to convert the generated the first order mode to the second order mode which has wider bandwidth than the LPFG converting fundamental mode to the second order mode.

The OAM mode generated mentioned above is only one pure OAM mode with the same polarization. There are other two types of mixed OAM modes composed of two orthogonal polarized OAM modes with the opposite TCs, according to the polarization state of the two orthogonal modes that can be generated in all-fiber system. One type is LP-OAM, while the other type is CP-OAM.

For the first type, the mixed OAM mode, composed of two orthogonal LP-OAM modes with the opposite TCs, can be obtained by the following equation:

$$E = \begin{cases} EH_{l-1,1}^{even} \pm iHE_{l+1,1}^{odd} = \hat{x}OAM_{\pm l} \pm i\hat{y}OAM_{\mp l} \\ EH_{l-1,1}^{odd} \mp iHE_{l+1,1}^{even} = \hat{x}OAM_{\mp l} \pm i\hat{y}OAM_{\pm l} \end{cases} \tag{8}$$

where E represents the mixed OAM mode. We can find that the mixed OAM modes are composed of x- and y-polarized LP-OAM modes with the opposite TCs, and these two orthogonal LP-OAM modes can be separated by a polarizer. In 2016, Jiang et al. experimentally demonstrated a method to generate optical vortices with tunable OAM in optical fiber by using a mechanical LPFG, a mechanical rotator and a metal parallel slab [34]. The tunable OAM mode can be seen as a combination of $HE_{l+1,m}^{even}$ ($HE_{l+1,m}^{odd}$) and $EH_{l-1,m}^{odd}$ ($EH_{l-1,m}^{even}$), with a $\pi/2$ phase shift. They experimentally achieved the smooth variation of OAM mode from $l = -1$ to $l = +1$ by adjusting the polarizer placed at the end of the fiber, as shown in Figure 8d.

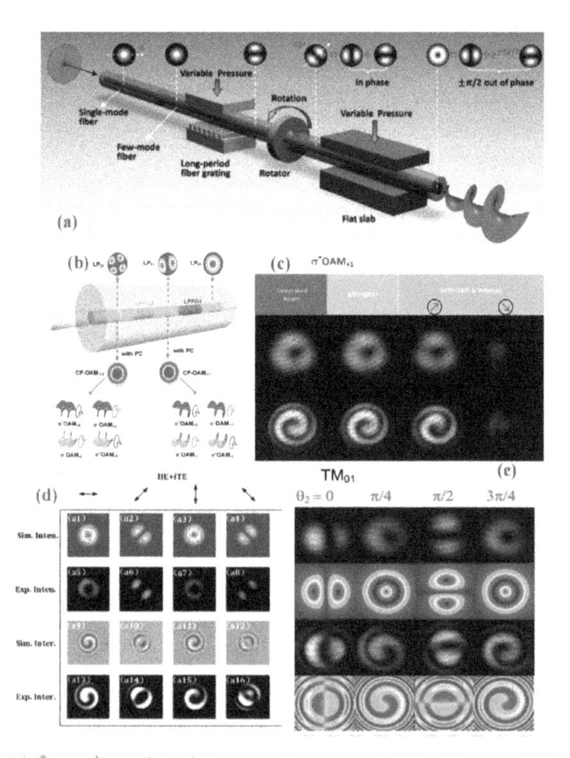

Figure 8. (**a**) and (**b**) Operating principle of OAM generation based on LPFGs system; (**c**)–(**e**) mode pattern of corresponding generated OAM modes in (**c**) [43], (**d**) [34], and (**e**) [32], respectively. Reprinted with permission from [35], copyright 2015 The Optical Society; [32], copyright 2018 The Optical society; [34], copyright 2016 The Optical Society; [43], copyright 2018 De Gruyter.

For the other type, the mixed OAM mode, composed of two orthogonal CP-OAM modes with the opposite TCs, can be obtained by the following equation:

$$\begin{aligned} EH^{even}_{l-1,m} &= \tfrac{1}{2}(\sigma^{-}OAM_{-l} + \sigma^{+}OAM_{+l}) \\ EH^{odd}_{l-1,m} &= -\tfrac{i}{2}(\sigma^{-}OAM_{-l} - \sigma^{+}OAM_{+l}) \\ HE^{even}_{l+1,m} &= \tfrac{1}{2}(\sigma^{+}OAM_{-l} + \sigma^{-}OAM_{+l}) \\ HE^{odd}_{l+1,m} &= \tfrac{i}{2}(\sigma^{+}OAM_{-l} - \sigma^{-}OAM_{+l}) \end{aligned} \quad (9)$$

We can see that the arbitrary single CV mode can be seen as a linear combination of a left-handed and a right-handed CP-OAM with the opposite TC. Since one of these two CP-OAM is left-handed, the other one is right-handed, we can use a QWP to make these two CP-OAM have an orthogonal projection. Then we can use a polarizer with optical axis rotating to one projection orientation to filter out the other CP-OAM. The authors of this paper have experimentally demonstrated that using a single CV mode in a two-mode fiber (TMF), the topological charge of generated OAM mode can be switched among −1, 0, +1, as shown in Figure 8e [32]. In our work, a CO_2-laser induced rocking LPFG inscribed in the two-mode fiber is fabricated to efficiently generate the CV mode, including TE_{01}, TM_{01} and $TE_{01} \pm TM_{01}$ mode. Then a QWP and a polarizer are used to separate the two CP-OAM modes.

The OAM modes obtained by [34,35,37,43] are all generated by a combination of two or four degenerated modes in the same mode group and the associated eigenmodes must maintain a stable phase and polarization. In addition, in ideal cylindrical waveguides such as fiber, the propagation constants of the TE_{01} and TM_{01} mode are different. For example, the effective index separation between the TE_{01} and TM_{01} mode is measured to be 6.6×10^{-4} in the fiber [36]. The purity of CP-OAM generated by combination of $TM_{01} \pm iTE_{01}$ or $HE_{21}^{even} \pm iTE_{01}$ cannot be maintained when the length of optical fiber is changed. However, the OAM modes generated by a single CV mode don't have this problem.

However, the generated OAM modes obtained by the studies mentioned above are dependent on the polarization state of the input light, namely, these all-fiber generators exhibited a sensitivity requiring specialized polarization states for the input light without adjusting the PC. Zhang et al. have proposed a polarization-independent OAM generator based on a chiral fiber grating (CFG) fabricated by twisting a fused few-mode fiber during hydrogen-oxygen flame heating [44]. They experimentally fabricate a left-handed CFG (LCFG) and a right-handed CFG (RCFG) to convert the fundamental mode to the OAM mode, and experimentally investigate the polarization characteristics, helical phase of the coupled mode in CFG for varying polarization state of input light. Their results showed that the coupled OAM mode had the same polarization state with the input fundamental mode. And the chirality of the generated OAM mode was polarization-independent and determined solely by the helicity of the CFG, as shown in Figure 9. The OAM mode generated by the RCFG can only be OAM_{+1} mode, while OAM mode generated by the LCFG can only be OAM_{-1} mode.

Compared to other mode converters, LPFGs have many advantages, such as low loss, small size, easy fabrication, and high coupling efficiency up to 99%, and so on. However, LPFGs also have many disadvantages compared with the mode selective coupler and the photonic lantern, for example, the narrow bandwidth and one LPFG can only convert the fundamental mode (the zeroth order mode) to a phase-matching higher order mode. Thus, there have been many attempts to increase the bandwidth of an LPFG, such as chirped LPFG [69], length-apodized LPFG [70], the cascading of several LPFGs [43], operating an LPFG at its turning point along its phase-matching curve [71], shortening the length of the LPFG, namely decreasing the number of grooves [72], and so on. However, chirping can significantly decrease the mode conversion efficiency and the bandwidth increase is still limited. For LPFG operating at its turning point, such a turning point may not exist for a given set of guided modes. For shortening the length of LPFG, it will decrease the coupling efficiency. For the cascading of several LPFGs, for example, the bandwidth of cascading two LPFGs that one LPFG convert fundamental mode to the first order mode and the other LPFG convert the first order mode to the second order mode is wider than the bandwidth of LPFG that directly converts fundamental mode to the second order mode [43], However, this method needs the fabrication of several LPFGs, which brings about extra loss.

4.2. Mode Selective Coupler (MSC)-Based OAM Generation Systems

Optical fiber couplers have been widely used in many research areas, such as optical communication, optical sensing, and fiber lasers. In addition, the fiber couplers can be fabricated from various types of fibers, such as single mode fiber, few mode fiber, polarization-maintaining fiber, and ring core fiber.

The operation principle of the MSC is shown in Figure 10. Composed of SMF and FMF, MSC is based on the phase-matching condition between the fundamental mode in the SMF and higher order modes in the FMF. This can be achieved by satisfying the ERI of the fundamental mode in the SMF equaling to that of the higher order mode in the FMF. When the phase-matching condition is satisfied, the fundamental mode in the SMF will be converted to the particular higher order mode in the FMF.

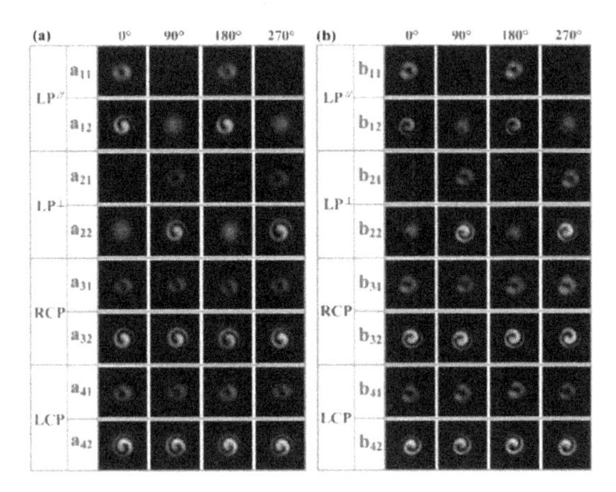

Figure 9. The mode pattern of generated OAM mode based on the (**a**) LCFG and (**b**) RCFG respectively; RCP: right circularly-polarized; LCP: left circularly-polarized. Reprinted with permission from [44], copyright 2019 The Optical Society.

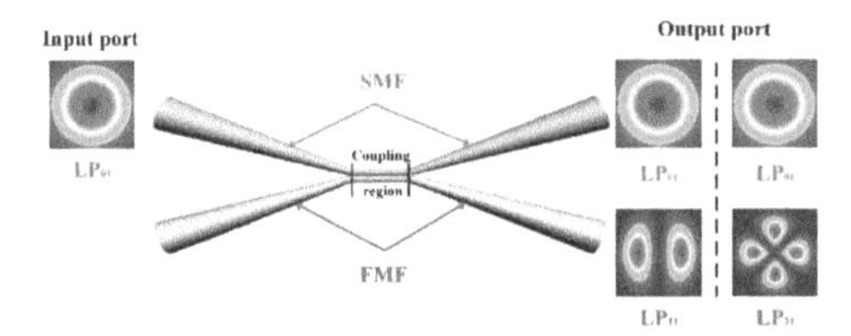

Figure 10. Schematics of the MSCs, composed of the SMF and the FMF. SMF: single-mode fiber; FMF: four-mode fiber.

In order to determine the coupling efficiency between the fundamental mode and high-order modes on the fiber tapering diameter, the following coupled equations are solved [73]:

$$\begin{aligned} \frac{dA_1(z)}{dz} &= i(\beta_1 + C_{11})A_1 + iC_{12}A_2 \\ \frac{dA_2(z)}{dz} &= i(\beta_2 + C_{22})A_2 + iC_{21}A_1 \end{aligned} \tag{10}$$

where z is the distance along the coupling region of the MSC, A_1 and A_2 are the slowly-varying field amplitudes in the SMF and FMF of the MSC, and β_1 and β_2 are the propagation constant of fundamental mode in the SMF and higher order mode to be coupled in the FMF, respectively. Due to the phase matching condition that ERI of fundamental mode in the SMF equaling to that of the first or second order modes in the FMF, β_1 should be equal to β_2. C_{11} and C_{22}, C_{12}, and C_{21} are the self-coupling and mutual coupling coefficients, respectively. Self-coupling coefficients are small relative to mutual coupling coefficients, and can be ignored. Moreover, the mutual coupling coefficients $C_{12} \approx C_{21} \approx C$, where C is a coefficient depending on the width and length of the coupling region. Thus, the power distribution in coupler can be given as follows [74]:

$$\begin{aligned} P_1(z) &= |A_1(z)|^2 = 1 - F^2\sin^2(\tfrac{C}{F}z) \\ P_2(z) &= F^2\sin^2(\tfrac{C}{F}z) \end{aligned} \tag{11}$$

where $F = \left(1 + \frac{\beta_1 - \beta_2}{4C^2}\right)^{-\frac{1}{2}}$, F^2 is the maximum coupling power between two fibers. According to Equation (11), it can be found that the power in coupling region exchanges periodically.

The MSC can be divided into two types according the fiber type used in its fabrication. One type is the MSC composed of the SMF and the FMF or multimode fiber (MMF). The FMF and MMF are similar to the SMF, which is composed of a core, cladding, and coating. The RI difference between the fiber core and cladding or the radius of the fiber core of FMF/MMF is larger than that of SMF. The other type is the MSC composed of the SMF and the ring core fiber (RCF), also called vortex fiber. The RCF is usually composed of three parts, namely, the fiber core, cladding, and ring. In these three parts, the RI of the ring is the highest. Thus, the light is able to be restricted and propagate in the ring. While it's invariant for the RI of the fiber core and cladding. The RCF can limit the radial order m to 1 and thereby fix the number of degenerated modes in each high azimuthal order mode group to be 4, which will decrease the multi-input-multi-output complexity. In addition, the RCF is more suitable for transmitting OAM modes than the FMF or MMF. However, the fabrication of RCFs is more difficult than the FMF/MMF. The fabrication of MSC, composed of SMF and the FMF/MMF is easier than the MSC composed of SMF and RCF, while the later MSC is more compatible to the OAM mode transmission system than the MSC composed of FMF/MMF. There have been many works to generate OAM mode based on the two types of MSCs [45–48]. The experimental setup usually used to generate the OAM mode based on the MSC is shown in Figure 11a The MSC is used to couple the fundamental mode to the higher order CV mode group, and the polarization controllers are used to redistribute the generated random state of lth-order CV modes to the OAM modes we want to obtain. The polarizer is used to identify the polarization state of the generated OAM modes or to separate the two orthogonal OAM modes.

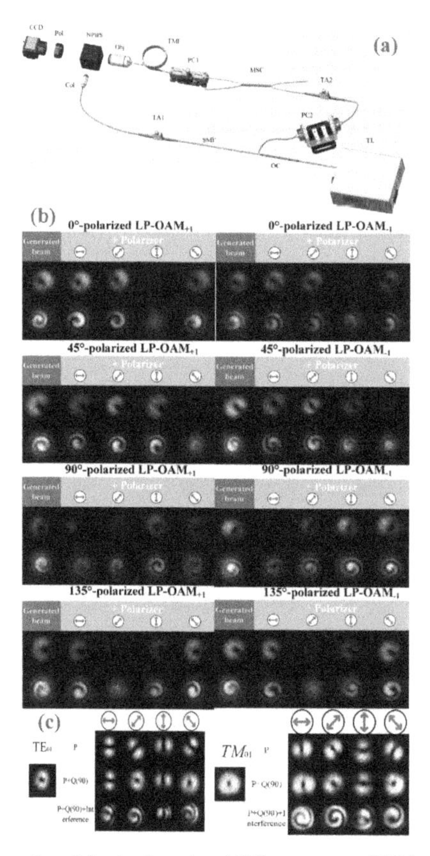

Figure 11. (**a**) A typical experimental setup based on MSC to generate OAM modes; (**b–c**) mode pattern of generated (**b**) LP-OAM, and (**c**) two orthogonal CP-OAM mode, respectively.

The authors of this paper have also experimentally demonstrated that arbitrary linearly-polarized OAM mode can be generated by carefully adjusting the PC without changing the polarization state of the input light based on the MSC, composed of a SMF and a TMF. The experimental results are shown in Figure 11b. We can find that the 0°-, 45°-, 90°-, and 135°-polarized LP-OAM modes with ±1 TC are generated successfully. In addition, we have also experimentally demonstrated the generations of $\sigma^{+}OAM_{\pm 1}$ and $\sigma^{-}OAM_{\pm 1}$ mode using a single first order CV mode, including TM_{01}, TE_{01}, HE_{21}^{even}, and HE_{21}^{odd} modes. These two CP-OAM modes can be separated by using a QWP and a polarizer with an angle $\pm\pi/4$. The results are shown in Figure 11c.

Yao et al. have proposed an all-fiber system to generate tunable first order OAM mode based on an all-fiber MSC, composed of an SMF and a TMF [48]. The MSC is used to couple fundamental mode to the first order mode. And they experimentally demonstrate the generation of first order OAM mode, produced by combining HE_{21}^{even} and TE_{01} (or HE_{21}^{odd} and TM_{01}) with a $\pi/2$ phase shift, and the topological charge of generated OAM mode can be tuned from -1 to +1 by adjusting the polarizer placed at the end of the FMF.

The excitation of first order CP-OAM modes obtained by Equation (7), using an MSC composed of an SMF and an RCF for the first time have been experimentally demonstrated in [47], and the mode purity of excited OAM mode is up to 75%.

Compared to the fiber grating-based mode converters mentioned above, the advantages of MSC include low loss, small size, easy fabrication, broad-bandwidth and the controllable coupling efficiency for the pure high-order mode which can be used in certain situations, such as in laser cavities, and so on. However, the MSCs also have disadvantages, for example, one MSC can only convert the fundamental mode to a particular high-order mode and, thus, we can only obtain one OAM mode by one MSC. If we want to multiplex N OAM modes, we need to cascade N MSCs which brings extra loss and complexity.

4.3. Micro-Structured Optical Fiber-Based OAM Generation System

The generation of OAM mode by the studies mentioned above are based on the all fiber system concluded in Figure 4. The mode coupling module can only couple the fundamental mode to the mixed mode composed of four degenerated modes in the lth-order CV mode group with random phase and amplitude. A field control module is used to redistribute the phase and amplitude of four degenerated modes to generate the OAM mode. It is usually difficult to obtain one particular OAM mode by adjusting the PC in the all-fiber system. It is more convenient if we can use a special fiber device to directly couple the fundamental mode to the OAM mode we want. There have been some reported works generating OAM modes directly through a special micro-structured optical fiber design [49–51].

There are two main operating principles of designing fiber for converting the fundamental mode to the high order OAM modes. One is based on the mode coupling theory. The designed fibers based on this principle are usually composed of one or several single-mode cores used to support the fundamental mode and one ring core used to support OAM mode. The electric fields of the fundamental core mode and the high order mode in the ring can be expressed as:

$$\begin{aligned} E_A(r) &= A(z)E_A(x,y)\exp(i\beta_A z) \\ E_B(r) &= B(z)E_B(x,y)\exp(i\beta_B z) \end{aligned} \tag{12}$$

where the coefficients $A(z)$ and $B(z)$ vary with z. And according to coupling mode theory, the two coefficients satisfy:

$$\begin{aligned} \frac{dA(z)}{dz} &= i\kappa_{AA}A + i\kappa_{AB}Be^{i(\beta_B-\beta_A)z} \\ \frac{dB(z)}{dz} &= i\kappa_{BB}B + i\kappa_{BA}Ae^{i(\beta_A-\beta_B)z} \end{aligned} \tag{13}$$

The mode coupling coefficient κ is given by:

$$\kappa_{\upsilon\mu} = \frac{\omega}{4}\int_{-\infty}^{\infty}\int_{-\infty}^{\infty} E_{\upsilon}^{*}(x,y)\cdot\Delta\varepsilon\cdot E_{\mu}(x,y)dxdy \tag{14}$$

where ω is the optical frequency, and $\triangle\varepsilon$ is the perturbation to the permittivity. The light power is supposed to only be launched into the fundamental mode core, so the coupling mode equation can be solved as:

$$\begin{aligned} A(z) &= (\cos\gamma z - \tfrac{i\delta}{\gamma}\sin\gamma z)e^{i\delta z} \\ B(z) &= \left(\tfrac{i\kappa_{BA}}{\gamma}\sin\gamma z\right)e^{-i\delta z} \end{aligned} \tag{15}$$

where $\gamma = (\kappa_{AB} \times \kappa_{BA} + \delta^2)^{1/2}$ and δ is the phase mismatching coefficient, where, $\delta = (\beta_B + \kappa_{BB} - \kappa_{AA} - \beta_A)/2$. The mode coupling efficiency can be expressed as:

$$\eta = \frac{|\kappa_{BA}|^2}{\gamma^2}\sin^2\gamma z \tag{16}$$

The authors of this paper have proposed and investigated a tunable microstructure optical fiber for different OAM mode generation by simulation based on the mode coupling theory. The microstructure optical fiber is composed of a high RI ring and a hollow core surrounded by four small air holes as shown in Figure 12 [49]. The hollow core and the surrounded four air holes are infiltrated by optical functional material whose RI can be modulated by physical parameters, leading to conversion between circularly-polarized fundamental mode and different OAM mode in the high RI ring with tunable operating wavelengths. The OAM modes are composed by $\hat{\sigma}^{\pm}OAM_{\pm l,1} = HE_{l+1,1}^{even} \pm iHE_{l+1,1}^{odd}$ and $\hat{\sigma}^{\mp}OAM_{\pm l,1} = EH_{l+1,1}^{even} \pm iEH_{l+1,1}^{odd}$, where l ranges from 2 to 8.

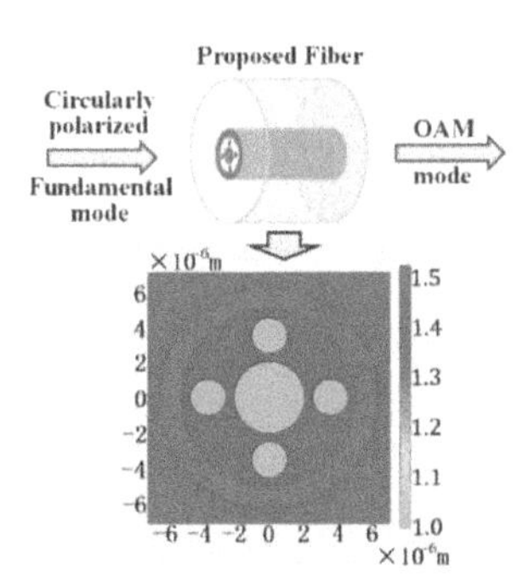

Figure 12. Schematic of proposed micro-structured optical fiber to generate OAM mode in [49]. Reprinted with permission from [49], copyright 2015 The Optical Society.

The other is to mimic the refractive spiral phase element [50,51]. This method typically uses the multi-core fiber (MCF) [50] or photonic crystal fiber (PCF) [51]. For the MCF, the phase change is a function of the core RI. Therefore, by arranging the RI distribution among the multi cores of the MCF. The phase difference between the adjacent core exactly equal $2\pi l/N$ where l is the TC of the desired OAM mode and N is the number of cores of MCF. With such a phase difference distribution, when the spatial-phase-modulated multi-beams converge in a section of ring core fiber (RCF), the OAM mode with TC l can be effectively generated [50]. For PCF, the air-holes in silica are arranged for the transverse subwavelength grating in [51]. The air-hole diameter across the transverse dimension can be varied to create an arbitrary ERI profile $N_{eff}(r, \theta) = N_0 + \triangle N_{eff}(r, \theta)$, where N_0 and $\triangle N_{eff}(r, \theta)$ are the constant and varying parts of $N_{eff}(r, \theta)$. To generate OAM with TC l, the introduced ERI perturbation $\triangle N_{eff}(r, \theta)$ must satisfy the phase-matching condition:

$$\Delta N_{eff}(r,\theta) = l\frac{\lambda}{z}\frac{\theta}{2\pi} \tag{17}$$

where z is the length of the PCF.

The OAM generation methods based on the fiber design not only have the advantages of the fiber device such as low loss and small size, but also have the advantages such as directly converting the fundamental mode to the OAM mode and the TCs of generated OAM being controllable. However, the structures of the designed fiber are usually complicated and difficult to fabricate in reality.

4.4. OAM Generation Based on Photonic Lantern

The photonic lantern (PL) is a low-loss optical device that connects several single-mode cores or few-mode cores to a single multimode core. Early interest in PLs were based on their original application in astronomical instrumentation [75,76], but recently its application in optical fiber communication has been attracting increased attention from researchers. The PL can be a spatial multiplexer for space-division multiplexing because one PL can multiplex N modes where N is the number of SMF used to fabricate the PL, which in principle allows the capacity of the communication system to be multiplied by N [77–89].

There are five types of PLs that have been reported to date [80–90]. The first type of PLs is fabricated by inserting N separate SMFs into a surrounding glass cane which is usually the PCFs made from glass with a pattern of air holes in the cladding [80,81]. Then the resulting glass body is heated and drawn down to form a taper transition to an MMF port. The MMF core is formed by the fused mass of SMFs, with the reduced-index cladding formed by the cane glass. This method has the advantages of easy control of the arrangement of SMFs by designing the arrangement of air holes of the cane. However, the SMFs are accommodated in separate compartments. The second type of PLs is fabricated by inserting a bundled N SMFs together into a capillary which has a lower RI than that of the SMF cladding [82,83]. Then the capillary is heated and drawn on a tapering machine to form the MMF port, which is more practical. However, the arrangement of SMFs with this method is more difficult than that of the first type of PLs, and with the number of SMFs increasing, the difficulty of fabricating PLs also increases. The third type of PLs is fabricated by inserting an MCF into a capillary which has a lower RI than that of the MCF cladding [84,85]. Then the capillary is heated and drawn on a tapering machine to form the MMF port. This method has the advantages, for example, several kilometers of the MSC can be drawn at once and this method needs only one MCF compared to the first and second type of PLs needing N SMFs. However, the MCF used in this method needs to be specially fabricated. The fourth type of PLs is fabricated by collapsing several holes in the multi-core PCF [86,87]. This is achieved by heating the PCF without significant stretching the fiber. Then the surface tension causes the holes to shrink and collapse completely which in effect makes the cores bigger until they merge to form a multimode core. The last type of PLs is fabricated by the direct laser writing technique on an integrated waveguide chip for locally modifying the structure of a substrate material in three dimensions [88–90]. There have been several studies regarding generating or multiplexing several OAM modes based on the mode-selective PL (MSPL) [52,53].

The generation of $OAM_{\pm 1} = LP_{11}^{a} \pm iLP_{11}^{b}$, $OAM_{\pm 2} = LP_{21}^{a} \pm iLP_{21}^{b}$, and the mixed OAM mode $OAM_{+1} + OAM_{-2}$ and $OAM_{-1} + OAM_{+2}$ are experimentally demonstrated by using the five-mode MSPL whose ERI profile is arranged to a ring shape in [52], as shown in Figure 13.

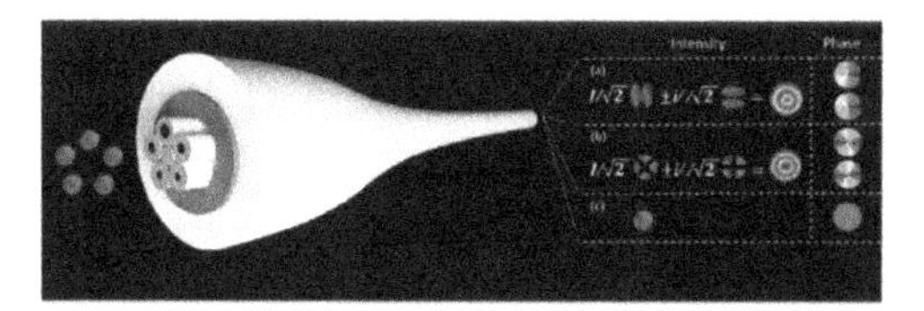

Figure 13. Schematic of operating principle of OAM generating based on the MSPL. Reprinted with permission from [52], copyright 2018 The Optical Society.

An all-fiber OAM multiplexer based on a six mode MSPL and a mode PC (MPC) that can multiplex both OAM modes of $-l$ and $+l$ up to the second order have been proposed in [53]. The generation of $OAM_{\pm 1}$, $OAM_{\pm 2}$, the mixed modes of arbitrary two OAM modes in these two OAM mode groups and the mixed mode of all four OAM modes have been experimentally demonstrated, as shown in Figure 14.

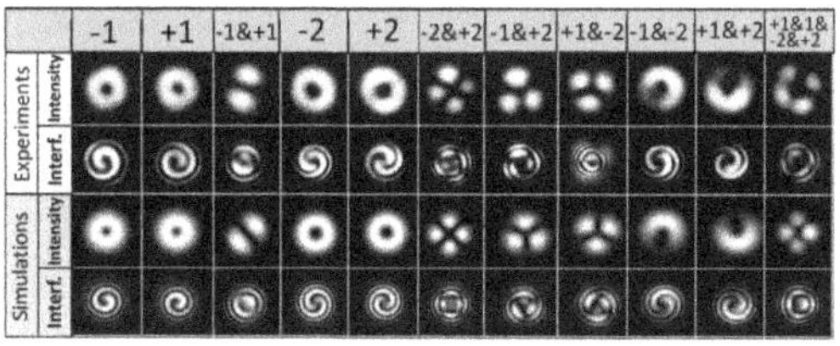

Figure 14. Mode pattern of generated OAM modes and the multiplexing of OAM modes using the MSPL. Reprinted with permission from [53], copyright 2018 The Optical Society.

Compared to the MSC and the LPFGs, the MSPL can multiplex N modes at the same time and, thus, it can be used to multiplex N OAM mode at the same time. However, with the number of SMF to fabricate the PL increasing, the difficulty of fabricating it also increases at the same time, and the insertion loss will also increase.

5. Discussions and Perspectives

As a new property of light that has been discovered relatively recently, OAM is playing an important role in various areas. There are two types of methods to generate OAM beams, spatial and fiber generating methods. Spatial generating methods have advantages in terms of flexible design and easily manipulation, but miniaturization is difficult. Compared with spatial methods, as natural azimuthal periodic beam-shapers, different types of fibers provide another way to generate OAM beams. Due to the miniaturization and low insertion loss, fiber methods occupy a place in OAM generation.

CV modes, OAM modes and LP modes are three types of fiber modes. Each possesses its own unique properties. Any electric field in the fiber is able to be decomposed into the superposition of one group of them, such as $E = \hat{x}OAM_{-l} = 0.5EH^{even}_{l-1,m} + 0.5iEH^{odd}_{l-1,m} + 0.5HE^{even}_{l+1,m} - 0.5iHE^{odd}_{l+1,m} = \hat{x}LP^{even}_{l,m} - i\hat{x}LP^{odd}_{l,m}$. Mathematically, they operate as vector bases describing the electric field in the fiber. A particular higher order spatial mode consists of four degenerated modes. The four degenerated modes can be CV modes, OAM modes or LP modes. They are able to completely describe lth order electric field in fiber, in their own forms. There exist transformation relations among these three modes. The so-called "OAM generation" is equivalent to adjusting the amplitudes and phases of the four degenerated modes in a particular order and simplifying the expression in OAM mode bases using physical methods. Taking $\hat{\sigma}^{+}OAM_{-l}$ as an example, if simultaneously calculating the equivalent CV modes, we obtain the typical expression similar to that found in almost all the studies, namely, $\hat{\sigma}^{+}OAM_{-1} = TM_{01} + iTE_{01}$. Equations (2)–(4) give all of these transformation relations among CV modes, OAM modes and LP modes. As for the detection of OAM beams, researchers have aimed to detect the TC, that is, the helical degree of the OAM beams. Because the physical nature is the phase detection, the interference method is mostly used. The TC of an OAM beam can be recognized from the interference patterns. The common interference patterns are vortex shape or fork-like shape, depending on the specific interference condition, namely, type of reference beam, divergence, tilted degree, and phase difference.

A number of devices can generate OAM beams in a fiber system, including the optical fiber grating, the mode selective coupler, the microstructure optical fiber and the photonics lantern. Among these devices, the optical fiber grating and the mode selective coupler are relatively mature. However, their defection is also obvious. They can just couple the fundamental mode to one particular higher order mode. The microstructure optical fiber seems to be better than the former. It has more freedom in design and may be potential to realize some unique functions that other devices can't access. However, the fabrication of microstructure fiber is still a significant problem. Many microstructural optical fibers only exist in theory and are difficult to fabricate. Photonic lantern is the most potential OAM generating device so far. PL can be regarded as an enhanced mode selective coupler, which is able to distribute core modes into different higher order modes in distinguishing channel. This brings a significant benefit in terms of demultiplexing. Additionally, if reversing the structure of the PL (which requires a re-design but not a direct reverse the demultiplexing PL), different higher order modes can merge into a single core. Multiplexing can also be realized. The technique of PL is not yet mature and still needs time to be optimized. The targets of improvement, such as the channel number, insertion loss, and crosstalk, are still far from applicable.

Although the physical processes of these devices are different, their fundamental principle is the same, that is, coupling the fundamental mode to higher order modes in the corresponding channels, and carefully adjusting the four degenerated modes in a particular higher order. In special combinations of amplitudes and phases among four degenerated modes, the OAM mode can be obtained. So far, two types of combination among the four degenerated modes can be used to generate the OAM mode. The first type is the pure OAM state. In this situation we can obtain an OAM beam directly. The other type is the state consisting of two orthogonal polarized OAM modes with the opposite TC. By selecting the polarization, we can obtain the corresponding OAM mode. The latter method exhibits an extra benefit that the TC is adjustable.

In summary, OAM beams are receiving increasing interest in various areas due to their novelty and potential applications. There are many methods used to generate OAM beams, each with its own advantages and disadvantages. With different bottlenecks, all the technique still require time to realize commercialization. As a branch of OAM generation methods, fiber methods have contributed to the generation of OAM beams with the advantages of miniaturization and low insertion loss but challenges in robustness. Despite this, along with the efforts from researchers all around the world, we may see an increasing number of applications based on OAM beams in the future.

Author Contributions: H.Z. mainly wrote the Section 1, Section 4, prepared the copyright and integrated the contents of advances. B.M. mainly wrote the Section 2, Section 3, Section 5. Y.H., Z.W. and Y.Y. provided constructive improvements and feedback. Y.L. largely guides the completion of this review.

Funding: This work was jointly supported by the National Natural Science Foundation of China under grant nos. 61835006, 11674177, 61775107 and 11704283, Tianjin Natural Science Foundation under grant no. 16JCZDJC31000, the National Key Research and Development Program of China under grant nos. 2018YFB0504401 and 2018YFB070, and the 111 Project (B16027).

Conflicts of Interest: The authors declare no conflict of interest.

References

1. Allen, L.; Beijersbergen, M.W.; Spreeuw, R.J.C.; Woerdman, J.P. Orbital angular momentum of light and the transformation of Laguerre-Gaussian laser modes. *Phys. Rev. A At. Mol. Opt. Phys.* **1992**, 45, 8185. [CrossRef]
2. Dholakia, K.; Cizmar, T. Shaping the future of manipulation. *Nat. Photonics* **2011**, 5, 335–342. [CrossRef]

3. Tkachenko, G.; Brasselet, E. Helicity-dependent three-dimensional optical trapping of chiral microparticles. *Nat. Commun.* **2014**, *5*, 4491. [CrossRef] [PubMed]
4. Paez-Lopez, R.; Ruiz, U.; Arrizon, V.; Ramos-Garcia, R. Optical manipulation using optimal annular vortices. *Opt. Lett.* **2016**, *41*, 4138–4141. [CrossRef] [PubMed]
5. Furhapter, S.; Jesacher, A.; Bernet, S.; Ritsch-Marte, M. Spiral interferometry. *Opt. Lett.* **2005**, *30*, 1953–1955. [CrossRef]
6. Curtis, J.E.; Koss, B.A.; Grier, D.G. Dynamic holographic optical tweezers. *Opt. Commun.* **2002**, *207*, 169–175. [CrossRef]
7. Curtis, J.E.; Grier, D.G. Modulated optical vortices. *Opt. Lett.* **2003**, *28*, 872–874. [CrossRef]
8. Padgett, M.; Bowman, R. Tweezers with a twist. *Nat. Photonics* **2011**, *5*, 343–348. [CrossRef]
9. Wang, J.; Yang, J.Y.; Fazal, I.M.; Ahmed, N.; Yan, Y.; Huang, H.; Ren, Y.X.; Yue, Y.; Dolinar, S.; Tur, M.; et al. Terabit free-space data transmission employing orbital angular momentum multiplexing. *Nat. Photonics* **2012**, *6*, 488–496. [CrossRef]
10. Yan, Y.; Xie, G.D.; Lavery, M.P.J.; Huang, H.; Ahmed, N.C.; Bao, J.; Ren, Y.X.; Cao, Y.W.; Li, L.; Zhao, Z.; et al. High-capacity millimetre-wave communications with orbital angular momentum multiplexing. *Nat. Commun.* **2014**, 5. [CrossRef]
11. Huang, H.; Milione, G.; Lavery, M.P.J.; Xie, G.D.; Ren, Y.X.; Cao, Y.W.; Ahmed, N.; Nguyen, T.A.; Nolan, D.A.; Li, M.J.; et al. Mode division multiplexing using an orbital angular momentum mode sorter and MIMO-DSP over a graded-index few-mode optical fibre. *Sci. Rep. UK* **2015**, *5*, 14931. [CrossRef] [PubMed]
12. Willner, A.E.; Huang, H.; Yan, Y.; Ren, Y.; Ahmed, N.; Xie, G.; Bao, C.; Li, L.; Cao, Y.; Zhao, Z.; et al. Optical communications using orbital angular momentum beams. *Adv. Opt. Photonics* **2015**, *7*, 66–106. [CrossRef]
13. Wang, J. Advances in communications using optical vortices. *Photonics Res.* **2016**, *4*, B14–B28. [CrossRef]
14. Li, Z.; Liu, W.; Li, Z.; Tang, C.; Cheng, H.; Li, J.; Chen, X.; Chen, S.; Tian, J. Tripling the capacity of optical vortices by nonlinear metasurface. *Laser Photonics Rev.* **2018**, 1800164. [CrossRef]
15. Yang, T.S.; Zhou, Z.Q.; Hua, Y.L.; Liu, X.; Li, Z.F.; Li, P.Y.; Ma, Y.; Liu, C.; Liang, P.J.; Li, X.; et al. Multiplexed storage and real-time manipulation based on a multiple degree-of-freedom quantum memory. *Nat. Commun.* **2018**, *9*, 3407. [CrossRef]
16. Gibson, G.; Courtial, J.; Padgett, M.J.; Vasnetsov, M.; Pas'ko, V.; Barnett, S.M.; Franke-Arnold, S. Free-space information transfer using light beams carrying orbital angular momentum. *Opt. Express* **2004**, *12*, 5448–5456. [CrossRef]
17. Beijersbergen, M.W.; Coerwinkel, R.P.C.; Kristensen, M.; Woerdman, J.P. Helical-wavefront laser beams produced with a spiral phaseplate. *Opt. Commun.* **1994**, *112*, 321–327. [CrossRef]
18. Uchida, M.; Tonomura, A. Generation of electron beams carrying orbital angular momentum. *Nature* **2010**, *464*, 737–739. [CrossRef]
19. Heckenberg, N.R.; Mcduff, R.; Smith, C.P.; White, A.G. Generation of optical phase singularities by computer-generated holograms. *Opt. Lett.* **1992**, *17*, 221. [CrossRef]
20. Mohammad, M.; Maga A-Loaiza, O.S.; Chang, C.; Brandon, R.; Mehul, M.; Boyd, R.W. Rapid generation of light beams carrying orbital angular momentum. *Opt. Express* **2013**, *21*, 30196–30203.
21. Yu, V.B.; Vasnetsov, M.V.; Soskin, M.S. Laser beams with screw dislocations in their wavefronts. *Nat. Genet.* **1990**, *47*, 73–77.
22. McMorran, B.J.; Agrawal, A.; Anderson, I.M.; Herzing, A.A.; Lezec, H.J.; McClelland, J.J.; Unguris, J. Electron vortex beams with high quanta of orbital angular momentum. *Science* **2011**, *331*, 192–195. [CrossRef]
23. Karimi, E.; Schulz, S.A.; Leon, I.D.; Qassim, H.; Upham, J.; Boyd, R.W. Generating optical orbital angular momentum at visible wavelengths using a plasmonic metasurface. *Light Sci. Appl.* **2014**, *3*, e167. [CrossRef]
24. Nanfang, Y.; Patrice, G.; Kats, M.A.; Francesco, A.; Jean-Philippe, T.; Federico, C.; Zeno, G. Light propagation with phase discontinuities: Generalized laws of reflection and refraction. *Science* **2011**, *334*, 333–337.
25. Zeng, J.; Wang, X.; Sun, J.; Pandey, A.; Cartwright, A.N.; Litchinitser, N.M. Manipulating complex light with metamaterials. *Sci. Rep.* **2013**, *3*, 2826. [CrossRef]
26. Zhe, Z.; Wang, J.; Willner, A.E. Metamaterials-based broadband generation of orbital angular momentum carrying vector beams. *Opt. Lett.* **2013**, *38*, 932–934. [CrossRef]
27. Beijersbergen, W.M.; Allen, V.D.; Veen, E.L.; Woerdman, O.H. Astigmatic laser mode converters and transfer of orbital angular momentum. *Opt. Commun.* **1993**, *96*, 123–132. [CrossRef]

28. Marrucci, L.; Karimi, E.; Slussarenko, S.; Piccirillo, B.; Santamato, E.; Nagali, E.; Sciarrino, F. Spin-to-orbital conversion of the angular momentum of light and its classical and quantum applications. *J. Opt.* **2011**, *13*, 064001. [CrossRef]
29. Oemrawsingh, S.S.R.; Houwelingen, J.A.W.; Van Eliel, E.R.; Woerdman, J.P.; Verstegen, E.J.K.; Kloosterboer, J.G.; Hooft, G.W. Production and characterization of spiral phase plates for optical wavelengths. *Appl. Opt.* **2004**, *43*, 688–694. [CrossRef]
30. Cai, X.L.; Wang, J.; Strain, M.J.; Morris, B.J.; Zhu, J.; Sorel, M.; O' Brien, J.L.; Thompson, M.G.; Yu, S. Integrated compact optical vortex beam emitters. *Science* **2012**, *338*, 363–366. [CrossRef]
31. Bozinovic, N.; Golowich, S.; Kristensen, P.; Ramachandran, S. Control of orbital angular momentum of light with optical fibers. *Opt. Lett.* **2012**, *37*, 2451–2453. [CrossRef] [PubMed]
32. Han, Y.; Chen, L.; Liu, Y.G.; Wang, Z.; Zhang, H.W.; Yang, K.; Chou, K.C. Orbital angular momentum transition of light using a cylindrical vector beam. *Opt. Lett.* **2018**, *43*, 2146–2149. [CrossRef] [PubMed]
33. Jiang, Y.C.; Ren, G.; Jin, W.X.; Xu, Y.; Jian, W.; Jian, S.S. Polarization properties of fiber-based orbital angular momentum modes. *Opt. Fiber Technol.* **2017**, *38*, 113–118. [CrossRef]
34. Jiang, Y.C.; Ren, G.B.; Lian, Y.D.; Zhu, B.F.; Jin, W.X.; Jian, S.S. Tunable orbital angular momentum generation in optical fibers. *Opt. Lett.* **2016**, *41*, 3535–3538. [CrossRef] [PubMed]
35. Li, S.H.; Mo, Q.; Hu, X.; Du, C.; Wang, J. Controllable all-fiber orbital angular momentum mode converter. *Opt. Lett.* **2015**, *40*, 4376–4379. [CrossRef]
36. Wang, L.; Vaity, P.; Ung, B.; Messaddeq, Y.; Rusch, L.A.; LaRochelle, S. Characterization of OAM fibers using fiber Bragg gratings. *Opt. Express* **2014**, *22*, 15653–15661. [CrossRef] [PubMed]
37. Wu, H.; Gao, S.C.; Huang, B.S.; Feng, Y.H.; Huang, X.C.; Liu, W.P.; Li, Z.H. All-fiber second-order optical vortex generation based on strong modulated long-period grating in a four-mode fiber. *Opt. Lett.* **2017**, *42*, 5210–5213. [CrossRef] [PubMed]
38. Wu, S.H.; Li, Y.; Feng, L.P.; Zeng, X.L.; Li, W.; Qiu, J.F.; Zuo, Y.; Hong, X.B.; Yu, H.; Chen, R.; et al. Continuously tunable orbital angular momentum generation controlled by input linear polarization. *Opt. Lett.* **2018**, *43*, 2130–2133. [CrossRef]
39. Zhang, X.Q.; Wang, A.T.; Chen, R.S.; Zhou, Y.; Ming, H.; Zhan, Q.W. Generation and conversion of higher order optical vortices in optical fiber with helical fiber Bragg gratings. *J. Lightwave Technol.* **2016**, *34*, 2413–2418. [CrossRef]
40. Zhao, Y.H.; Liu, Y.Q.; Zhang, C.Y.; Zhang, L.; Zheng, G.J.; Mou, C.B.; Wen, J.X.; Wang, T.Y. All-fiber mode converter based on long-period fiber gratings written in few-mode fiber. *Opt. Lett.* **2017**, *42*, 4708–4711. [CrossRef] [PubMed]
41. Jiang, Y.C.; Ren, G.B.; Li, H.S.; Tang, M.; Jin, W.X.; Jian, W.; Jian, S.S. Tunable orbital angular momentum generation based on two orthogonal LP modes in optical fibers. *IEEE Photonics Technol. Lett.* **2017**, *29*, 901–904. [CrossRef]
42. Li, Y.J.; Jin, L.; Wu, H.; Gao, S.C.; Feng, Y.H.; Li, Z.H. Superposing multiple LP Modes with microphase difference distributed along fiber to generate OAM mode. *IEEE Photonics J.* **2017**, 9. [CrossRef]
43. Han, Y.; Liu, Y.G.; Wang, Z.; Huang, W.; Chen, L.; Zhang, H.W.; Yang, K. Controllable all-fiber generation/conversionof circularly polarized orbital angular momentum beams using long period fiber gratings. *Nanophotonics* **2018**, *7*, 287–293. [CrossRef]
44. Zhang, Y.; Bai, Z.Y.; Fu, C.L.; Liu, S.; Tang, J.; Yu, J.; Liao, C.R.; Wang, Y.; He, J.; Wang, Y.P. Polarization-independent orbital angular momentum generator based on a chiral fiber grating. *Opt. Lett.* **2019**, *44*, 61–64. [CrossRef]
45. Heng, X.B.; Gan, J.L.; Zhang, Z.S.; Li, J.; Li, M.Q.; Zhao, H.; Qian, Q.; Xu, S.H.; Yang, Z. M All-fiber stable orbital angular momentum beam Generation and propagation. *Opt. Express* **2018**, *26*, 17429–17436. [CrossRef] [PubMed]
46. Jiang, Y.C.; Ren, G.B.; Shen, Y.; Xu, Y.; Jin, W.X.; Wu, Y.; Jian, W.; Jian, S.S. Two-dimensional tunable orbital angular momentum generation using a vortex fiber. *Opt. Lett.* **2017**, *42*, 5014–5017. [CrossRef]
47. Pidishety, S.; Pachava, S.; Gregg, P.; Ramachandran, S.; Brambilla, G.; Srinivasan, B. Orbital angular momentum beam excitation using an all-fiber weakly fused mode selective coupler. *Opt. Lett.* **2017**, *42*, 4347–4350. [CrossRef]
48. Yao, S.Z.; Ren, G.B.; Shen, Y.; Jiang, Y.C.; Zhu, B.F.; Jian, S.S. Tunable orbital angular momentum generation using all-fiber fused coupler. *IEEE Photonics Technol. Lett.* **2018**, *30*, 99–102. [CrossRef]

49. Huang, W.; Liu, Y.G.; Wang, Z.; Zhang, W.C.; Luo, M.M.; Liu, X.Q.; Guo, J.Q.; Liu, B.; Lin, L. Generation and excitation of different orbital angular momentum states in a tunable microstructure optical fiber. *Opt. Express* **2015**, *23*, 33741–33752. [CrossRef]
50. Seghilani, M.; Azana, J. All-Fiber OAM generation/conversion using helically patterned photonic crystal fiber. *IEEE Photonics Technol. Lett.* **2018**, *30*, 347–350. [CrossRef]
51. Heng, X.B.; Gan, J.L.; Zhang, Z.S.; Qian, Q.; Xu, S.H.; Yang, Z.M. All-fiber orbital angular momentum mode generation and transmission system. *Opt. Commun.* **2017**, *403*, 180–184. [CrossRef]
52. Eznaveh, Z.S.; Zacarlas, J.C.A.; Lopez, J.E.A.; Shi, K.; Milione, G.; Jung, Y.M.; Thomsen, B.C.; Richardson, D.J.; Fontaine, N.; Leon-Saval, S.G.; et al. Photonic lantern broadband orbital angular momentum mode multiplexer. *Opt. Express* **2018**, *26*, 30042–30051. [CrossRef] [PubMed]
53. Zeng, X.L.; Lin, Y.; Feng, L.P.; Wu, S.H.; Yang, C.; Li, W.; Tong, W.J.; Wu, J. All-fiber orbital angular momentum mode multiplexer based on a mode-selective photonic lantern and a mode polarization controller. *Opt. Lett.* **2018**, *43*, 4779–4782. [CrossRef] [PubMed]
54. Mao, B.W.; Liu, Y.G.; Zhang, H.W.; Yang, K.; Han, Y.; Wang, Z.; Li, Z.H. Complex analysis between CV modes and OAM modes in fiber systems. *Nanophotonics* **2018**. [CrossRef]
55. Kisała, P.; Harasim, D.; Mroczka, J. Temperature-insensitive simultaneous rotation and displacement (bending) sensor based on tilted fiber Bragg grating. *Opt. Express* **2016**, *24*, 29922–29929. [CrossRef] [PubMed]
56. Gao, Y.; Sun, J.Q.; Chen, G.D.; Sima, C. Demonstration of simultaneous mode conversion and demultiplexing for mode and wavelength division multiplexing systems based on tilted few-mode fiber Bragg gratings. *Opt. Express* **2015**, *23*, 9959–9967. [CrossRef] [PubMed]
57. Jin, L.; Wang, Z.; Liu, Y.; Kai, G.; Dong, X. Ultraviolet-inscribed long period gratings in all-solid photonic bandgap fibers. *Opt. Express* **2008**, *16*, 21119–21131. [CrossRef]
58. Liao, C.; Wang, Y.; Wang, D.N.; Jin, L. Femtosecond laser inscribed long-period gratings in all-solid photonic bandgap fibers. *IEEE Photonics Technol. Lett.* **2010**, *22*, 425–427. [CrossRef]
59. Rao, Y.J.; Wang, Y.P.; Ran, Z.L.; Zhu, T. Novel fiber-optic sensors based on long-period fiber gratings written by high-frequency CO_2 laser pulses. *J. Lightwave Technol.* **2003**, *21*, 1320–1327.
60. Zhang, X.H.; Liu, Y.G.; Wang, Z.; Yu, J.; Zhang, H.W. LP01-LP11a mode converters based on long-period fiber gratings in a two-mode polarization-maintaining photonic crystal fiber. *Opt. Express* **2018**, *26*, 7013–7021. [CrossRef]
61. Wang, Y. Review of long period fiber gratings written by CO_2 laser. *J. Appl. Phys.* **2018**, *108*, 081101-1–081101-08. [CrossRef]
62. Bai, Z.Y.; Zahng, W.G.; Gao, S.C.; Geng, P.C.; Zhang, H.; Li, J.L.; Liu, F. Compact long period fiber grating based on periodic micro-core-offset. *IEEE Photon. Technol. Lett.* **2013**, *25*, 2111–2113. [CrossRef]
63. Yin, G.L.; Wang, Y.P.; Liao, C.R.; Zhou, J.T.; Zhong, X.Y.; Wang, G.J.; Sun, B.; He, J. Long period fiber gratings inscribed by periodically tapering a fiber. *IEEE Photonics Technol. Lett.* **2014**, *26*, 698–701. [CrossRef]
64. Diez, A.; Birks, T.A.; Reeves, W.H.; Mangan, B.J.; Russell, P.S.J. Excitation of cladding modes in photonic crystal fibers by flexural acoustic waves. *Opt. Lett.* **2000**, *25*, 1499–1501. [CrossRef]
65. Savin, S.; Digonnet, M.J.F.; Kino, G.S.; Shaw, H.J. Tunable mechanically induced long-period fiber gratings. *Opt. Lett.* **2000**, *25*, 710–712. [CrossRef]
66. Shen, X.; Hu, X.W.; Yang, L.Y.; Dai, N.L.; Wu, J.J.; Zhang, F.F.; Peng, J.G.; Li, H.Q.; Li, J.Y. Helical long-period grating manufactured with a CO2 laser on multicore fiber. *Opt. Express* **2017**, *25*, 10405–10412.
67. Huang, W.; Liu, Y.G.; Wang, Z.; Liu, B.; Wang, J.; Luo, M.M.; Guo, J.Q.; Lin, L. Multi-component-intermodalinterference mechanism and characteristics of a long period grating assistant fluid-filled photonic crystal fiber interferometer. *Opt. Express* **2014**, *22*, 5883–5894. [CrossRef]
68. Zhou, Q.; Zhang, W.; Chen, L.; Bai, Z.; Zhang, L.; Wang, L.; Wang, B.; Yan, T. Bending vector sensor based on a sector-shaped long-period grating. *IEEE Photonics Technol. Lett.* **2015**, *27*, 713–716. [CrossRef]
69. Liu, Q.; Chiang, K.S. Design of long-period waveguide grating filter by control of waveguide cladding profile. *J. Lightwave Technol.* **2006**, *24*, 3540–3546.
70. Wang, W.; Wu, J.Y.; Chen, K.X.; Jin, W.; Chiang, K.S. Ultra-broadband mode converters based on length-apodized long-period waveguide gratings. *Opt. Express* **2017**, *25*, 14341–14350. [CrossRef]
71. Liu, Q.; Chiang, K.S.; Lor, K.P. Dual resonance in a long-period waveguide grating. *Appl. Phys. B* **2007**, *86*, 147–150. [CrossRef]

72. Ostling, D.; Engan, H.E. Broadband spatial mode conversion by chirped fiber bending. *Opt. Lett.* **1996**, *21*, 19–24. [CrossRef]
73. Wang, T.; Wang, F.; Shi, F.; Pang, F.F.; Huang, S.J.; Wang, T.Y.; Zeng, X.L. Generation of femtosecond optical vortex beams in all-fiber mode-locked fiber laser using mode selective coupler. *J. Lightwave Technol.* **2017**, *35*, 2161–2166. [CrossRef]
74. Xiao, Y.L.; Liu, Y.G.; Wang, Z.; Liu, X.Q.; Luo, M.M. Design and experimental study of mode selective all-fiber fused mode coupler based on few mode fiber. *Acta Phys. Sin.* **2015**, *64*, 204207.
75. Leon-Saval, S.G.; Birks, T.A.; Bland-Hawthorn, J.; Englund, M. Single-Mode Performance in Multimode Fibre Devices. In Proceedings of the Optical Fiber Communication Conference, Anaheim, CA, USA, 6 March 2005.
76. Leon-Saval, S.G.; Birks, T.A.; Bland-Hawthorn, J.; Englund, M. Multimode fiber devices with single-mode performance. *Opt. Lett.* **2005**, *30*, 2545–2547. [CrossRef]
77. Bland-Hawthorn, J.; Kern, P. Molding the flow of light: Photonics in astronomy. *Phys. Today* **2012**, *65*, 31–37. [CrossRef]
78. Thomson, R.R.; Harris, R.J.; Birks, T.A.; Brown, G.; Allington-Smith, J.; Bland-Hawthorn, J. Ultrafast laser inscription of a 121-waveguide fan-out for astrophotonics. *Opt. Lett.* **2012**, *37*, 2331–2333. [CrossRef]
79. Birks, T.A.; Mangan, B.J.; Díez, A.; Cruz, J.L.; Murphy, D.F. 'Photonic lantern' spectral filters in multi-core fibre. *Opt. Express* **2012**, *20*, 13996–14008. [CrossRef]
80. Yerolatsitis, S.; Birks, T.A. Tapered Mode Multiplexers Based on Standard Single-Mode Fibre. In Proceedings of the European Conference on Optical Communication, London, UK, 22–26 September 2013. paper PD1.C.1.
81. Fontaine, N.K.; Leon-Saval, S.G.; Ryf, R.; Salazar Gil, J.R.; Ercan, B.; Bland-Hawthorn, J. Mode-Selective Dissimilar Fiber Photonic-Lantern Spatial Multiplexers for Few-Mode Fiber. In Proceedings of the European Conference on Optical Communication, London, UK, 22–26 September 2013. paper PD1.C.3.
82. Noordegraaf, D.; Skovgaard, P.M.W.; Nielsen, M.D.; Bland-Hawthorn, J. Efficient multi-mode to single-mode coupling in a photonic lantern. *Opt. Express* **2009**, *17*, 1988–1994. [CrossRef]
83. Noordegraaf, D.; Skovgaard, P.M.W.; Maack, M.D.; Bland-Hawthorn, J.; Haynes, R.; Lægsgaard, J. Multi-mode to single-mode conversion in a 61 port photonic lantern. *Opt. Express* **2010**, *18*, 4673–4678. [CrossRef]
84. Thomson, R.R.; Brown, G.; Kar, A.K.; Birks, T.A.; Bland-Hawthorn, J. An Integrated Fan-Out Device for Astrophotonics. In Proceedings of the Frontiers in Optics, OSA Annual Meeting, Rochester, NY, USA, 24–28 October 2010. paper PDPA3.
85. Birks, T.A.; Mangan, B.J.; Díez, A.; Cruz, J.L.; Leon-Saval, S.G.; Bland-Hawthorn, J.; Murphy, D.F. Multicore Optical Fibres for Astrophotonics. In Proceedings of the European Conference on Lasers and Electro-Optics, Munich, Germany, 22–26 May 2011. paper JSIII2.1.
86. Yerolatsitis, S.; Birks, T.A. Three-Mode Multiplexer in Photonic Crystal Fibre. In Proceedings of the European Conference on Optical Communication, London, UK, 22–26 September 2013. paper Mo.4.A.4.
87. Yerolatsitis, S.; Gris-Sánchez, I.; Birks, T.A. Adiabatically-tapered fiber mode multiplexers. *Opt. Express* **2014**, *22*, 608–617. [CrossRef]
88. Said, A.A.; Dugan, M.; Bado, P.; Bellouard, Y.; Scott, A.; Mabesa, J. Manufacturing by laser direct-write of three-dimensional devices containing optical and microfluidic networks. *Proc. SPIE* **2004**, *5339*, 194–204.
89. Nasu, Y.; Kohtoku, M.; Hibino, Y. Low-loss waveguides written with a femtosecond laser for flexible interconnection in a planar light-wave circuit. *Opt. Lett.* **2005**, *30*, 723–725. [CrossRef]
90. Arriola, A.; Gross, S.; Jovanovic, N.; Charles, N.; Tuthill, P.G.; Olaizola, S.M.; Fuerbach, A.; Withford, M.J. Low bend loss waveguides enable compact, efficient 3D photonic chips. *Opt. Express* **2013**, *21*, 2978–2986. [CrossRef]

Review

A Review of Tunable Orbital Angular Momentum Modes in Fiber: Principle and Generation

Lipeng Feng, Yan Li *, Sihan Wu, Wei Li, Jifang Qiu, Hongxiang Guo, Xiaobin Hong, Yong Zuo and Jian Wu *

The State Key Laboratory of Information Photonics and Optical Communications, Beijing University of Posts and Telecommunications, Beijing 100876, China; 15201017279@bupt.edu.cn (L.F.); sihanwu@bupt.edu.cn (S.W.); w_li@bupt.edu.cn (W.L.); jifangqiu@bupt.edu.cn (J.Q.); hxguo@bupt.edu.cn (H.G.); xbhong@bupt.edu.cn (X.H.); yong_zuo@bupt.edu.cn (Y.Z.)

* Correspondence: liyan1980@bupt.edu.cn (Y.L.); jianwu@bupt.edu.cn (J.W.)

Received: 1 April 2019; Accepted: 17 May 2019; Published: 13 June 2019

Abstract: Orbital angular momentum (OAM) beams, a new fundamental degree of freedom, have excited a great diversity of interest due to a variety of emerging applications. The scalability of OAM has always been a topic of discussion because it plays an important role in many applications, such as expanding to large capacity and adjusting the trapped particle rotation speed. Thus, the generation of arbitrary tunable OAM mode has been paid increasing attention. In this paper, the basic concepts of classical OAM modes are introduced firstly. Then, the tunable OAM modes are categorized into three types according to the orbital angular momentums and polarization states of mode carrying. In order to understand the OAM evolution of a mode intuitively, three kinds of Poincaré spheres (PSs) are introduced to represent the three kinds of tunable OAM modes. Numerous methods generating tunable OAM modes can be roughly divided into two types: spatial and fiber-based generation methods. The principles of fiber-based generation methods are interpreted by introducing two mode bases (linearly-polarized modes and vector modes) of the fiber. Finally, the strengths and weaknesses of each generation method are pointed out and the key challenges for tunable OAM modes are discussed.

Keywords: orbital angular momentum; tunable OAM; Poincaré sphere; state of polarization

1. Introduction

A light beam has two "rotational" degrees of freedom: spin angular momentum (SAM) and orbital angular momentum (OAM) [1]. The SAM per photon is $\sigma\hbar$ (where $\hbar$ is the Plank's constant h divided by 2π), which is related to the state of polarization for left-circular $\sigma = +1$, for right-circular $\sigma = -1$, while for linearly polarized light $\sigma = 0$. For elliptically polarized light, the SAM varies from zero to $\pm\ 1\hbar$ as the state of polarization varies from linear to circular. The OAM is associated to the phase structure of the complex electric field with a helical phase front defined by the factor of $\exp(il\theta)$, which carry a definite amount of OAM per photon equal to lh. As the OAM beams have a phase singularity, they have a doughnut-shaped spatial profile with zero intensity at the center. Considerable interest in orbital angular momentum arises over its potential applications in multiple fields. For instance, orbital angular momentum can be transferred to trapped suitable material particles causing them to rotate, which enables optical manipulation, trapping and tweezers in fields as diverse as biosciences and micromechanics [2–5]; the angular momentum of light can be used to encode quantum information that is carried by the corresponding photon states [6]; the exploitation of the orbital angular momentum also opens the door to the generation and manipulation of multi-dimensional quantum entangled states with an arbitrarily large number of entanglement dimensions [7–10]; the doughnut intensity profile and phase singularity of OAM contribute to contrast enhancement techniques by depleting the fluorescence everywhere except at the dark center of the depletion beam, which enables resolution microscopy beyond the diffraction limit [11–13]; the OAM modes are also applied in optical communications (fiber and free space) with large capacity and long ranges due to its partial robustness against turbulence [14,15]; and astrophysical processes may generate photonic OAM, such as light scattering off inhomogeneities in the environments surrounding energetic sources (masers, pulsars, quasars) and light scattering off rotating black holes [16], etc.

The unique advantage of using optical OAM in these applications relies, to a large extent, on the use of multiple different OAM states. For example, the multiple available OAM states facilitate high-dimensional quantum information processing and large-capacity optical communications. The rotation speed of the trapped micro-particles for optical micro-manipulation is also related to the states of OAM and required to be continuously adjusted [17,18]. Thus, the control of the orbital angular momentum in a beam is important.

In this paper, an overview of the basic concepts and generation methods of the tunable OAM modes is given. Firstly, the classical OAM mode is introduced in the Section 2, including the physical concept, mathematical expression and generation methods. Secondly, the basic concepts and theoretical expressions of three kinds of tunable OAM modes are briefly described in Section 3. In order to intuitively understand the OAM evolution, three Poincaré spheres (PSs) are used to represent the three kinds of tunable OAM modes, which is similar to the polarization PS. In addition, the relationships among three PSs are concluded. Then, the fiber-based and free-space generation methods are respectively classified into three types according to the controllable variables in the Section 4. Finally, the advantages and disadvantages of each generation method are listed and the key challenges for tunable OAM modes are discussed in Section 5.

2. The Classical Orbital Angular Momentum (OAM) Mode

The light beams carrying orbital angular momentum are spatially structured beams with helical phase fronts. For the points on the mode cross-section with the same radius, the polarization states and amplitudes are the same, but with different phases. The electromagnetic field of an classical OAM beam is identified by a phase term expressed as $\exp(\pm il\theta)$, where θ is the azimuthal angle in the transverse plane of the mode. The l which called as the topological charge means the number of 2π phase shifts along the circle around the beam axis [1,19]. The sign of l is relative to the handedness of helical phase front. The positive is for left helical phase front and negative is for right helical phase front (from the point of view of the receiver). In principle, l can take an arbitrary integer number ranged from $-\infty$ to $+\infty$, therefore, the state of OAM-carrying mode is infinite. Figure 1 shows the helical phase fronts of $l = 0, 1, -1$ and 2. Meanwhile, the OAM is the component of angular momentum of a light beam that is only dependent on the spatial field distribution but not on the polarization. Thus, the OAM mode can be classified into two types according to the state of polarization (SOP) that the beam carries. One type is the linearly-polarized OAM (LP-OAM) whose polarization states of every point on the mode cross-section are the linear polarization. This kind of OAM mode has no SAM. The other type is circularly-polarized OAM (CP-OAM) whose polarization states of every point on the mode cross-section are the circular polarization.

In the free-space system, the OAM beams can be generated via numerous methods such as spatial light modulators [20], computer-generated fork holograms [21], spiral phase plates [22], cylindrical lens pairs [23], q-plates [24], etc. Although those spatial generation methods have the advantages of strong scalability and low crosstalk, there are two common defections for them, i.e., high insertion loss and large volume.

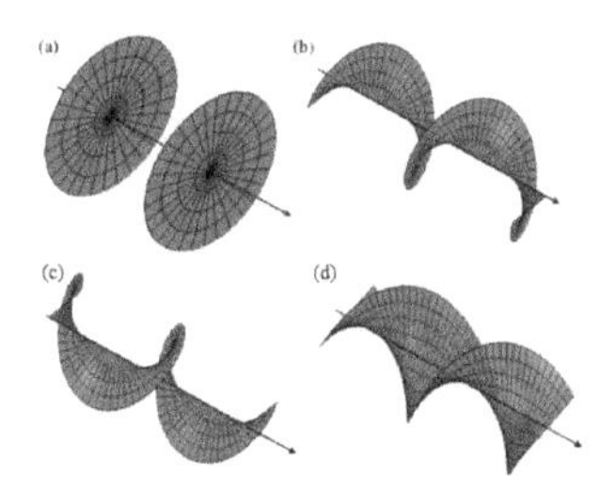

Figure 1. Helical phase fronts for (**a**) $l = 0$, (**b**) $l = 1$, (**c**) $l = -1$, and (**d**) $l = 2$. Reprinted with permission from [19], Copyright 2011 The Optical Society of America.

Compared with free-space generation systems, fiber-based methods have the advantage of the low insertion loss. Moreover, the devices in fiber-based generation methods are smaller, which greatly facilitates miniaturization. In the fiber, the $OAM_{\pm l}$ modes can be obtained by combining the two sets of fiber mode bases, LP mode and vector mode bases. The LP mode bases (LP_{lmax}, LP_{lmay}, LP_{lmbx} and LP_{lmby}) and vector mode bases ($HE^{e/o}$, $EH^{e/o}$, TE and TM) are respectively scalar and vector solutions to Maxwell Equation in the fiber [25]. Here, "e" and "o" refer to the even and odd modes, "m" is the radial order, denoting the number of radial nodes of the mode and l is the azimuthal order. The l order LP mode bases can be expressed by the Equation (1).

$$\begin{aligned} LP_{lmax} &= \vec{x}F_{lm}(r)\cos(l\theta)LP_{lmay} = \vec{y}F_{lm}(r)\cos(l\theta) \\ LP_{lmbx} &= \vec{x}F_{lm}(r)\sin(l\theta)LP_{lmby} = \vec{y}F_{lm}(r)\sin(l\theta) \end{aligned} \tag{1}$$

where $\vec{x}$ and $\vec{y}$ denote the linear polarization along the x-axis and y-axis, respectively; $F_{lm}(r)$ represents the radial field distribution and θ is the azimuthal coordinate. The vector mode bases have the following transverse electric field distributions:

$$\begin{aligned} &\left\{ \begin{array}{c} HE^{e}_{l+1,m} \\ HE^{o}_{l+1,m} \end{array} \right\} = F_{l,m}(r)\left\{ \begin{array}{c} \vec{x}\cos(l\theta) - \vec{y}\sin(l\theta) \\ \vec{x}\sin(l\theta) + \vec{y}\cos(l\theta) \end{array} \right\} \quad (l \geq 1) \\ &\left\{ \begin{array}{c} EH^{e}_{l-1,m} \\ EH^{o}_{l-1,m} \end{array} \right\} = F_{l,m}(r)\left\{ \begin{array}{c} \vec{x}\cos(l\theta) + \vec{y}\sin(l\theta) \\ \vec{x}\sin(l\theta) - \vec{y}\cos(l\theta) \end{array} \right\} \quad (l > 1) \\ &\left\{ \begin{array}{c} TM_{0,m} \\ TE_{0,m} \end{array} \right\} = F_{l,m}(r)\left\{ \begin{array}{c} \vec{x}\cos(\theta) + \vec{y}\sin(\theta) \\ \vec{x}\sin(\theta) - \vec{y}\cos(\theta) \end{array} \right\} \quad (l = 1) \end{aligned} \tag{2}$$

When two LP modes owning the same polarization directions are combined with a $\pm\pi/2$ phase shift, the LP-OAM mode with same polarization as the LP modes is generated [26], as shown in Equation (3); When the even and odd modes of same vector mode (HE or EH) are combined with $\pm\pi/2$ phase shift, the superimposed mode is CP-OAM mode, as shown in Equation (4). The OAM handedness and SAM handedness of modes based on the HE bases are the same, while those of modes based on the EH bases are opposite [25].

$$\left\{ \begin{array}{c} LP_{lmax} \pm iLP_{lmbx} \\ LP_{lmay} \pm iLP_{lmby} \end{array} \right\} = F_{lm}(r)\left\{ \begin{array}{c} \vec{x}OAM_{\pm l,m} \\ \vec{y}OAM_{\pm l,m} \end{array} \right\} \tag{3}$$

$$\left\{ \begin{array}{cc} HE^{e}_{l+1,m} \pm iHE^{o}_{l+1,m} & (l > 1) \\ EH^{e}_{l-1,m} \pm iEH^{o}_{l-1,m} & (l > 1) \end{array} \right\} \text{or} \left\{ \begin{array}{cc} HE^{e}_{l+1,m} \pm iHE^{o}_{l+1,m} & (l = 1) \\ EH^{e}_{l-1,m} \pm iEH^{o}_{l-1,m} & (l = 1) \end{array} \right\} = F_{l,m}(r)\left\{ \begin{array}{cc} \sigma^{\pm}OAM_{\pm l,m} & (l \geq 1) \\ \sigma^{\mp}OAM_{\pm l,m} & (l \geq 1) \end{array} \right\} \tag{4}$$

3. Three Kinds of Tunable OAM Modes

3.1. The OAM Varies from −l to l with Homogeneous State of Polarization (SOP) along the Longitude of Orbital Poincaré Sphere (PS)

The mode with OAM from $-l$ to l and homogeneous SOP can be produced by overlapping two collinear classical OAM modes with variable relative amplitudes and equal but opposite phase chirality [16]. This kind of tunable OAM mode could maintain a constant geometry and total intensity during tuning, which is valuable for some applications such as the vortex tweezers and optical manipulation [2–5,17,18]. It is worth noting that the polarizations of two collinear classical OAM modes are the same. The superimposed mode is described by the equation:

$$\psi_1 = a_1|\mathrm{N}_l\rangle + b_1 e^{i\varphi_1}|\mathrm{S}_l\rangle \tag{5}$$

with positive, real amplitudes a_1, b_1 and relative phase ϕ_1. In addition, the squares of amplitudes add up to unity. N_l and S_l are the two OAM modes owning topological charge $-l$ and $+l$ with the same polarization, respectively.

$$\begin{aligned} |\mathrm{N}_l\rangle &= e^{-il\theta}(\mathrm{E_x}\vec{x} + \mathrm{E_y}\vec{y}) / \sqrt{(|\mathrm{E_x}|2 + |\mathrm{E_y}|2)} \\ |\mathrm{S}_l\rangle &= e^{+il\theta}(\mathrm{E_x}\vec{x} + \mathrm{E_y}\vec{y}) / \sqrt{(|\mathrm{E_x}|2 + |\mathrm{E_y}|2)} \end{aligned} \tag{6}$$

where $\mathrm{E_x}$ and $\mathrm{E_y}$ are the complex amplitudes of x and y polarizations, respectively. In agreement with intuitive arguments, the amplitudes a_1 and b_1 govern the relative contribution of $\mathrm{OAM}_{+\ell}$ and $\mathrm{OAM}_{-\ell}$ to the local and total orbital angular momentum. In general, the average OAM value that mode carries is calculated from the power in each OAM mode, as shown in Equation (7). The P_l represents the power in each OAM mode [27].

$$\mathrm{L_{ave}} = \frac{\sum lP_l}{\sum P_l} = \frac{l \times |a|^2 + (-l) \times |b|^2}{|a|^2 + |b|^2} \tag{7}$$

In 1999, M. J. Padgett and J. Courtial proposed an orbital PS to represent this kind of tunable OAM mode intuitively [28]. The north and south poles of the orbital PS represent the OAM modes with equal ℓ value but opposite helicity, respectively. Similar to the polarization PS [29], all the points on the orbital PS can be described as the superposition of the two poles. Therefore, according to Equation (5), the orbital PS can be used to describe completely all the states for this kind of tunable OAM. Figure 2a,b show the mode patterns, phase distributions and polarization states on the two orbital PSs when $l = 1$ and $l = 4$, respectively.

Unlike the classical OAM mode, the amplitude and phase of the superimposed mode vary with azimuthal coordinate, which can be changed by modulating the relative amplitudes of OAM_{+l} and $\mathrm{OAM}_{-\ell}$ modes. When a = 0 or b = 0, the superimposed mode has the phase of a classical OAM_{+l} or $\mathrm{OAM}_{-\ell}$ mode which lies on the south or north pole, as the S_l and N_l points shown in Figure 2a,b; When a = b, the phase of the mode is binary with $2l$ alternating phase segments of 0 and π, which is equivalent to the phase of the optical cogwheel. This mode carries no orbital angular momentum and possesses $2l$ intensity peaks about the azimuthal coordinate, which is the $\mathrm{LP}_{\ell\mathrm{m}}$ mode of fiber and lies on the equator. The intensity fringes will occur to rotate about the center of the resultant beam by adjusting the relative phase (ϕ_1) of the two overlapping modes. We show the mode patterns of two points (H_l and V_l) which are located at the start and end points of the diameter on the equator. The H_l and V_l points are the combination mode of two poles owning same amplitude with $\phi_1 = 0$ and π, respectively.

$$|\mathrm{H}_l\rangle = \frac{|\mathrm{N}_l\rangle + |\mathrm{S}_l\rangle}{\sqrt{2(|\mathrm{E_x}|2+|\mathrm{E_y}|2)}} = \frac{\sqrt{2}[\cos(l\theta)((\mathrm{E_x}\vec{\mathrm{x}} + \mathrm{E_y}\vec{\mathrm{y}})]}{\sqrt{(|\mathrm{E_x}|2+|\mathrm{E_y}|2)}}$$
$$|\mathrm{V}_l\rangle = -i\frac{|\mathrm{N}_l\rangle - |\mathrm{S}_l\rangle}{\sqrt{2(|\mathrm{E_x}|2+|\mathrm{E_y}|2)}} = \frac{-\sqrt{2}[\sin(l\theta)((\mathrm{E_x}\vec{\mathrm{x}} + \mathrm{E_y}\vec{\mathrm{y}})]}{\sqrt{(|\mathrm{E_x}|2+|\mathrm{E_y}|2)}} \quad (8)$$

When $0 < |a_1 - b_1| < 1$, the local curvature of the helical wavefront is no longer constant nor linear. However, the phase singularity remains, qualifying the beam as a kind of optical vortex. Meanwhile, the intensity distribution also becomes an intermediate state between the LP mode and OAM mode. For $0 < a_1 - b_1 < 1$, the superimposed mode has the negative average OAM value and hence lies on the upper hemisphere. For $-1 < a_1 - b_1 < 0$, the superimposed mode has the positive average OAM value and lies on the lower hemisphere. The A_l and B_l show two specific points on the upper and lower hemisphere, respectively. Therefore, when the relative amplitude $(a_1 - b_1)$ varies from −1 to 1, the superimposed mode will change along the longitude and the average OAM value will also vary from l to $-l$.

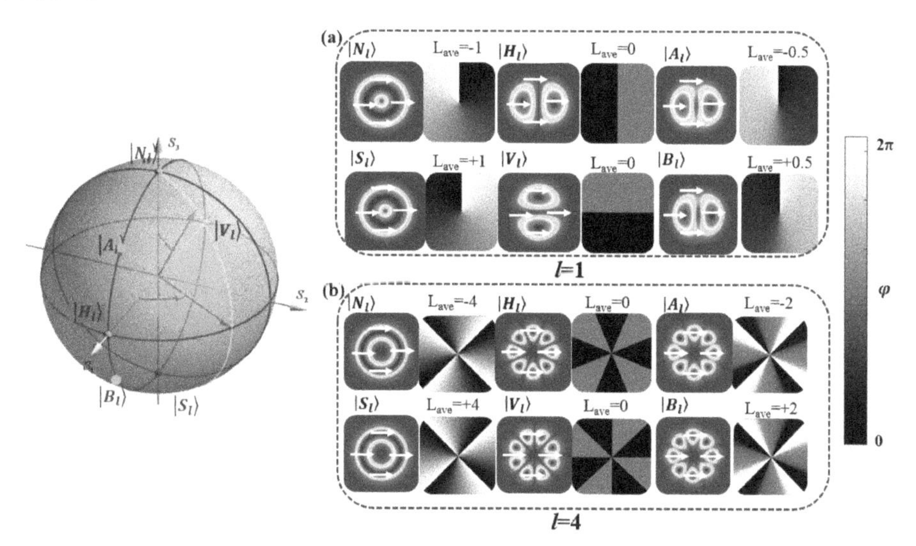

Figure 2. (**a**) Intensity profiles, polarization states and the phase distributions of superimposed mode when $l = 1$ and (**b**) $l = 4$. The north pole N_l and south pole S_l represent orthogonal circularly polarized modes with topological charges of $-l$ and $+l$; The points H_l and V_l represent the two points on the equator; The A_l and B_l points of (**a**) represent the two modes with average orbital angular momentum (OAM) values of −0.5 and 0.5 for $l = 1$. The A_l and B_l points of (**b**) represent the two modes with average OAM values of −2 and 2 for $l = 4$.

3.2. The OAM Varies from −l to l with Inhomogeneous SOP Along the Longitude of Higher-Order PS

The above tunable OAM modes have conventional homogeneous polarizations. In other words, the SOP of each point in the mode is same and invariant along the azimuthal coordinate. Recently there has been increasing interest in the modes with inhomogeneous SOPs. For those vector vortex (VV) modes, each point of the electrical field is the same polarization (linear, elliptical and circular polarization), but the polarized direction of each point is related to the azimuthal coordinate, such as the radial and azimuthal polarized cylindrical vector (CV) beams [30]. The VV modes extend the properties of conventional homogeneous polarization, such as the ability to produce strong longitudinal field components and smaller waist sizes upon focusing by high numerical aperture objectives, which may have important applications in nanoscale optical imaging and manipulation [31–35].

A higher-order PS is introduced as the theoretical framework for describing the spatially inhomogeneous SOPs of generalized vortex modes [36,37], as shown in Figure 3. Similar to the points on the orbital PS, arbitrary ones on the higher-order PS can be obtained by the linear combination of the modes on the two poles. The two poles are orthogonal CP-OAM modes with opposite topological charge.

$$\psi_2 = a_2\left|N_l^R\right\rangle + b_2 e^{i\varphi_2}\left|S_l^L\right\rangle \tag{9}$$

where

$$\left|N_l^R\right\rangle = e^{-il\theta}(\vec{x} + i\vec{y})/\sqrt{2} \tag{10}$$

$$\left|S_l^L\right\rangle = e^{+il\theta}(\vec{x} - i\vec{y})/\sqrt{2} \tag{11}$$

Equations (10) and (11) represent right and left circularly-polarized OAM modes with topological charge $-l$ and $+l$, respectively. The coefficients a_2 and b_2 are the amplitudes of the Equations (10) and (11), respectively, and the ϕ_2 is the relative phase between them. The higher-order PS has five salient features: (1) For $l > 1$, the OAM and SAM handedness of each pole can be the same or opposite, therefore two spheres are needed to describe higher-order SOPs of VV modes. (2) All the modes on the PS have annular intensity profiles and a dark hollow center, which possess phase or polarization singularities. (3) The modes can degenerate to the modes on the orbital PS through a linear polarizer, e.g., horizontally orientated as depicted by the double-sided arrows in the Figure 3a,b. (4) When the state of mode changes along the longitude, the average OAM value varies from $-l$ to l and the SAM changes from -1 to 1 (1 to -1), correspondingly. Figure 3a,b show intensity profiles, polarization states and phase distributions of six points on the two higher-order PSs with $l = \pm1$, respectively. For $l = +1$, the handedness of OAM and SAM on each pole is opposite, as N_l^R and S_l^L show in Figure 3a. This higher-order PS can completely characterize a general cylindrical vector mode [38], such as radial and azimuthal polarization, which are equivalent to TE and TM fiber modes. The H_l and V_l points are the TE_{01} and TM_{01} fiber modes which are obtained by the combination mode of two poles owning same amplitudes with $\phi_2 = 0$ and π, respectively. The deduction processes about H_l and V_l points are described by Equation (12). In addition, we choose two specific points (A_l and B_l) to illustrate the intensity and polarization distributions on the upper and lower hemisphere with -0.5 and 0.5 average OAM values, respectively.

$$\begin{aligned} |H_l\rangle &= \frac{\left|N_l^R\right\rangle + \left|S_l^L\right\rangle}{2} = \cos(l\theta)\vec{x} + \sin(l\theta)\vec{y} \\ |V_l\rangle &= -i\frac{\left|N_l^R\right\rangle - \left|S_l^L\right\rangle}{2} = -\sin(l\theta)\vec{x} + \cos(l\theta)\vec{y} \end{aligned} \tag{12}$$

For $l = -1$, the handedness of OAM and that of SAM on each pole are the same, shown in the Figure 3b. Similarly, N_l^R and S_l^L represent the modes on the north and south poles. This higher-order PS can describe the so-called π-vector modes [39] which are equivalent to the HE_{21}^e and HE_{21}^o fiber modes. H_l and V_l are the two points on the equator, whose average OAM values are both 0. Intermediate modes between the pole and equator have the elliptical polarizations and annular intensity profiles. The A_l and B_l are, respectively, the points with -0.5 and 0.5 average OAM values.

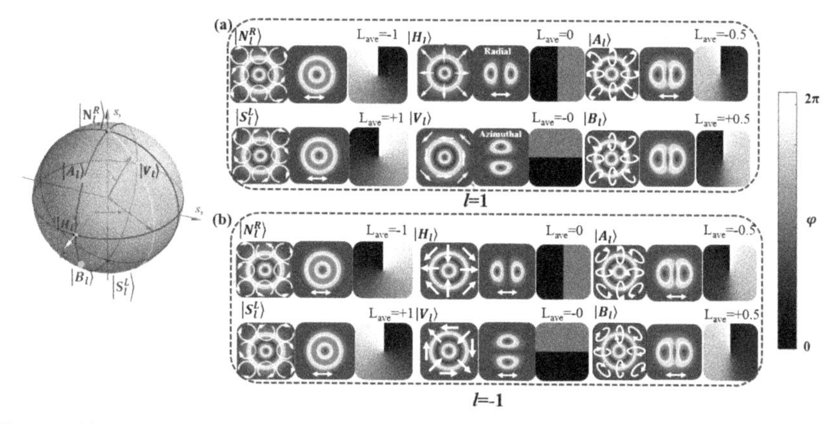

Figure 3. (**a**) Intensity profiles, polarization states and phase distributions of superimposed modes when $l = 1$ and (**b**) $l = -1$. The north pole N_l^R and south pole S_l^L represent orthogonal circularly-polarized modes with topological charges of $-l$ and $+l$; The points H_l and V_l represent the two points on the equator; The A_l and B_l points represent the two modes with −0.5 and 0.5 average OAM values, respectively.

3.3. The OAM Varies from l to n with Inhomogeneous SOP Along the Longitude of Hybrid-Order PS

For the second kind of tunable OAM mode, the polarization states and OAMs of modes on the higher-order PS are still confined to some special cases. For example, the modes on the equator have the azimuthally and radially linear polarization, but they are only the single vector beams possessing spatially inhomogeneous SOP and carrying zero-order OAM. The OAM modes on the poles of higher-order PS are only the single vortex beams with spiral wavefronts. Compared to a single vector mode and a single vortex mode, a vector vortex mode provides more degrees of freedom in optical manipulation [40,41]. Hence, in 2015, the hybrid-order PS is proposed to describe the evolution of the OAM and SOP, which extends the orbital PS and higher-order PS to a more general form [42].

The representation of the modes on the hybrid-order PS is the same as those on the higher-order PS except the orbital states on the poles. The orbital states of the two poles on orbital and higher-order PSs have the same value but opposite signs. Unlike the previous PSs, the orbital states of the poles on the hybrid-order PS are not confined to the same order topological charge and can be chosen arbitrarily. Any one mode on the hybrid-order PS can be expressed as the superposition of two poles:

$$\psi_3 = a_3\left|N_l^R\right\rangle + b_3 e^{i\varphi_3}\left|S_n^L\right\rangle \tag{13}$$

$$\left|N_l^R\right\rangle = e^{il\theta}(\vec{x} + i\vec{y})/\sqrt{2} \tag{14}$$

$$\left|S_n^L\right\rangle = e^{in\theta}(\vec{x} - i\vec{y})/\sqrt{2} \tag{15}$$

Equations (14) and (15) represent right and left circularly-polarized modes with different topological charges l and n, respectively. Any mode on the hybrid-order PS can be achieved by changing the coefficients a_3 and $b_3e^{i\varphi_3}$. Generally, the equatorial points on the hybrid-order PS represent the superposition of the two poles with equal intensities. The $H_{l,n}$ and $V_{l,n}$ of Figure 4 are the two equatorial points when $\phi_3 = 0$ and π, respectively.

$$\begin{aligned}\left|H_{l,n}\right\rangle &= \frac{\left|N_l^L\right\rangle + \left|S_n^L\right\rangle}{2} = \exp\frac{i(l+n)\theta}{2}\left[\cos\frac{(l-n)\theta}{2}\vec{x} + \sin\frac{(l-n)\theta}{2}\vec{y}\right] \\ \left|V_{l,n}\right\rangle &= -i\frac{\left|N_l^L\right\rangle - \left|S_n^L\right\rangle}{2} = \exp\frac{i(l+n)\theta}{2}\left[\cos\left(\frac{l-n}{2}\theta + \frac{\pi}{2}\right)\vec{x} + \sin\left(\frac{l-n}{2}\theta + \frac{\pi}{2}\right)\vec{y}\right]\end{aligned} \tag{16}$$

From Equation (16), it should be noted that the equatorial points represent modes carrying $(l + n)/2$ per photon. The relative phase of the superposition determines the orientation of the longitude.

Figure 4 depicts a hybrid-order PS at the situation of the north pole with state $\sigma = +1$ and $l = 0$, while the south pole with $\sigma = -1$ and $n = +2$. The N_l^R and S_n^L separately represent north and south poles; The $H_{l,n}$ and $V_{l,n}$ indicate two points on the equator. The $A_{l,n}$ and $B_{l,n}$ denote two points on the upper and lower hemispheres with 0.5 and 1.5 average OAM values. The average OAM that modes carry will change from l to $(l + n)/2$ and then to n along the longitude on the hybrid-order PS.

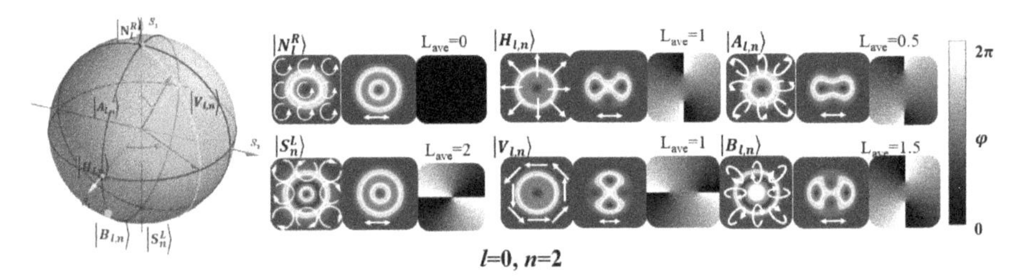

Figure 4. Intensity profiles, polarization states and phase distributions of superimposed modes when $\ell = 1$ and $n = 2$. The north pole N_l^R and south pole S_n^L represent orthogonal circularly-polarized modes with topological charges of l and n; The points $H_{l,n}$ and $V_{l,n}$ represent the two points on the equator; The $A_{l,n}$ and $B_{l,n}$ points represent the two modes with 0.5 and 1.5 average OAM values, respectively.

3.4. The Relationship Among the Three Kinds of Tunable OAM Modes

According to the representations of three kinds of tunable OAM modes, we find that there is a progressive relationship from orbital PS to higher-order PS, then to hybrid PS. Figure 5 shows the intuitive sketch describing the phase and polarization distributions of north and south poles on the three PSs. The white arrows represent polarization states and black-and-white images represent phase distributions. From the orbital PS to higher-order PS, the SOPs on the two poles vary from the same polarization to orthogonal circularly-polarized polarization. From the higher-order PS to hybrid-order PS, the SOPs on the two poles are kept while the orbital states vary from same order to different orders.

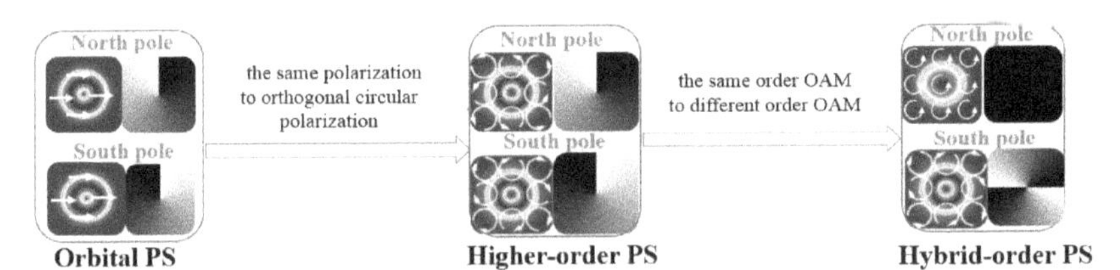

Figure 5. Sketch of the progressive relationship from the orbital Poincaré sphere (PS) to higher-order PS, then to hybrid-order PS.

4. Methods for Generation of Tunable OAM

The three kinds of tunable OAM modes can be considered as not only the superposition of classical OAM modes physically, but also the combination of x and y polarizations with spatially-variant amplitude and phase mathematically. Thus, the modes can be realized by some free-spatially optical elements that change the polarization of each point on the cross section of mode. Equations (17)–(19) show purely mathematical expressions in the form of Jones vectors, where the first and second elements of the vectors represent components of the field along the horizontal (x) and vertical (y) axes. The $A(\theta)$, $B(\theta)$, $\varphi_x(\theta)$ and $\varphi_y(\theta)$ represent the spatially-variant amplitude and phase factors, respectively.

$$\text{Orbital PS}: \begin{bmatrix} E_x(a_1 e^{-il\theta} + b_1 e^{i\varphi_1} e^{+il\theta}) \\ E_y(a_1 e^{-il\theta} + b_1 e^{i\varphi_1} e^{+il\theta}) \end{bmatrix} = \begin{bmatrix} E_x(A(\theta) e^{i\varphi_x(\theta)}) \\ E_y(B(\theta) e^{i\varphi_y(\theta)}) \end{bmatrix} \tag{17}$$

$$\text{Higher-order PS}: \begin{bmatrix} (a_2e^{-il\theta} + b_2e^{i\varphi_2}e^{+il\theta}) \\ i(a_2e^{-il\theta} - b_2e^{i\varphi_2}e^{+il\theta}) \end{bmatrix} = \begin{bmatrix} A(\theta)e^{i\varphi_x(\theta)} \\ B(\theta)e^{i\varphi_y(\theta)} \end{bmatrix} \tag{18}$$

$$\text{Hybrid-order PS}: e^{i\frac{(l+n)}{2}\theta} \begin{bmatrix} (a_3e^{i\frac{(l-n)}{2}\theta} + b_3e^{i\varphi_3}e^{-i\frac{(l-n)}{2}\theta}) \\ i(a_3e^{i\frac{(l-n)}{2}\theta} - b_3e^{i\varphi_3}e^{-i\frac{(l-n)}{2}\theta}) \end{bmatrix} = e^{i\frac{(l+n)}{2}\theta} \begin{bmatrix} A(\theta)e^{i\varphi_x(\theta)} \\ B(\theta)e^{i\varphi_y(\theta)} \end{bmatrix} \tag{19}$$

4.1. Free Space Method for Generation of Tunable OAM

Free-space generation methods of tunable OAM modes are generally assisted by spatial light modulators (SLMs) [17,43–51], deformable mirrors (DMs) [51], q-plate cells [52–55] and metasurfaces [56], and spiral phase plates (SPPs) [22]. Those optical elements can change the phase distribution of a mode, where the SLM and DM are programmable and can control the phase dynamically. In addition, the DM and SPP are polarization insensitive.

SLM is a computer-addressable reflective liquid crystal (LC) display which can impose any desired phase profile onto an incoming collimated beam by controlling the voltage (V) of each SLM pixel [46]. The phase retardation for each SLM pixel can be described as a function of the voltage (V) applied: $\delta(V) = (2\pi/\lambda)(n_e(V) - n_o)d$, where d is the thickness of the LC layer, n_e and n_o are the extraordinary and ordinary refractive indices of the LC retarder, respectively. Because of the birefringent nature of LC, when the input polarization state makes a projection on both the fast and slow axes of the SLM, the polarization state can be altered. The polarization property of the SLM can be exploited by the appropriate optical setup to achieve the desired change in the polarization. Moreover, the combination of the SLM and wave plates can be used to control the amplitudes of x and y polarizations due to the birefringent nature. For the SLM, it has the advantage of high flexibility due to the arbitrarily adjustable phase distribution, but also has the disadvantages of maximum power density limitation and large loss.

The DM is composed of many units. Each unit has its own independent controller. Under the control of external voltage, it can transform the wavefront phase [51]. In principle, the SLM modulates the wave-front phase by controlling the refractive index, and the DM modulates the wave-front by changing the distance of light propagation. As a phase controller, the DM is energy efficient and highly flexible, while the range of controllable phase is limited.

The q-plate cell is essentially birefringent waveplates with a uniform birefringent phase retardation δ across the plate thickness (which can be electrically controlled) and a space-variant transverse optical axis distribution exhibiting a topological charge "2q" [52]. The charge "q" represents the number of rotations of the local optical axis in a path circling once around the center of the plate. When the q-plate cell is illuminated by a circularly-polarized vortex beam, the output beam from the q-plate cell is the combination of two different-order OAM modes with adjustable amplitudes. For the q-plate usually used, it is a q-plate cell with a uniform birefringent phase retardation $\delta = \pi$. If a Gaussian mode with arbitrary polarization passes through the q-plate, the output mode will perform the following linear transformation, as shown in Equation (20) [54], where σ^+ and σ^- respectively indicate the left and right circular polarization. Thus, a q-plate cell can convert a Gauss beam to a vector beam and generate a vortex phase in one step. However, the "q" value is fixed, so the flexibility is poor.

$$q\cdot(A\sigma^+ + B\sigma^-) = A\sigma^- e^{i2q\theta} + B\sigma^+ e^{-i2q\theta} \tag{20}$$

The metasurface with tailorable structure geometry, as a two-dimensional electromagnetic nanostructure, possesses unparalleled advantages in optical phase and polarization manipulation, especially in subwavelength scale [57]. The operation principle is the same as the q-plate. But it is the difficult to fabricate and untunable once fabricated.

Free-space generation methods can be classified into three types according to the controllable variables. Figures 6–8 simply show schematic diagrams used to generate the tunable OAM mode based on the free-space system. The components marked by red and gray frames in Figures 6–8 are the adjusted ones and fixed ones, respectively.

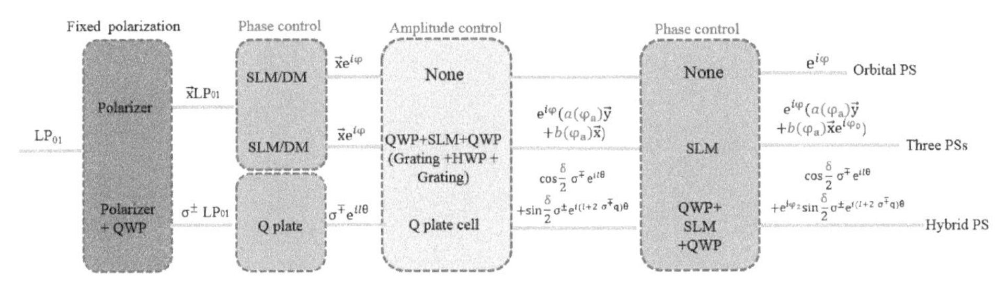

Figure 6. The flow chart of tunable OAM generation by adjusting the phase distributions on the spatial light modulators (SLMs), deformable mirrors (DMs) and q-plate cells.

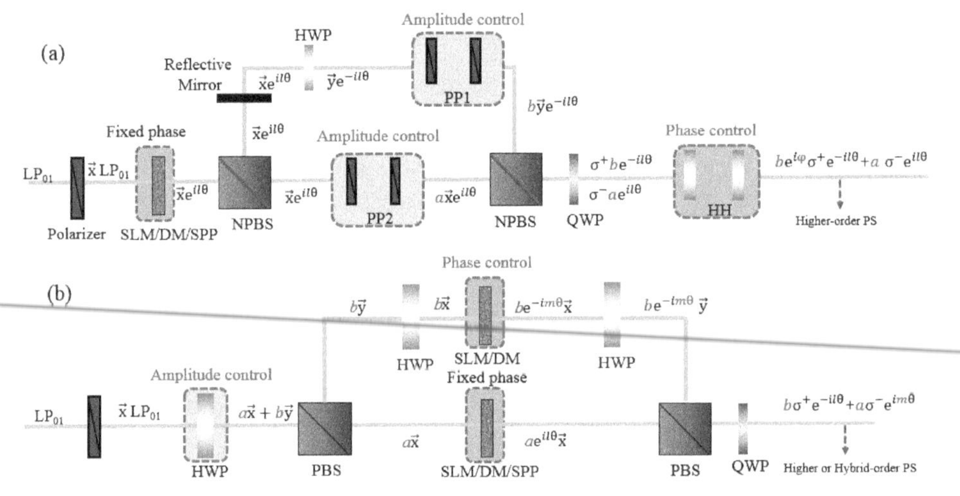

Figure 7. The flow chart of tunable OAM generation by interfering with two Laguerre-Gaussian modes. Adjusting phase and amplitude factors of two branches by (**a**) optical elements and (**b**) a HWP and a SLM.

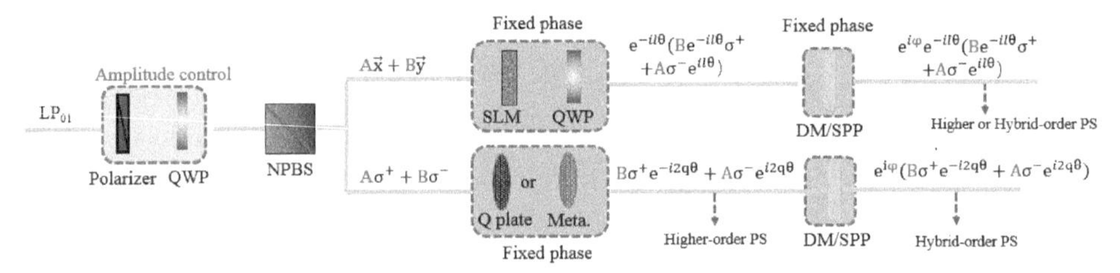

Figure 8. The flow chart of tunable OAM generation by continuously changing the state of input polarization.

The first kind of generation method is shown in Figure 6, which obtains tunable OAM modes by changing the phase information imposed on the SLMs, DMs or q-plate cells. When a Gaussian beam with linear polarization is launched to a programmable phase element (SLM or DM), the topological charge can be tuned by electrically changing the phase information loaded on them, which can obtain the modes on the orbital PS, as shown the first row in Figure 6 [17,43–46,52]. The second row of Figure 6b shows 3 sections as subsystems to control respectively three degrees of freedom in the optical field, e.g., the phase, amplitude and retardation between the x and y components [47,48]. The section of amplitude is achieved by the combination of two quarter-wave plates (QWPs) and a SLM [46] or the combination of a half-wave plate (HWP) and two diffraction gratings [47]. In ref. [46], the fast axes of the QWPs are along 45° and 135° with respect to the horizontal axis, respectively. By loading appropriate phase information on the three sections, this system can generate arbitrary modes on the three PSs. The modes on the three PSs can be obtained when the mathematical expression of the output mode equals to the one of Equations (17)–(19). When a Gaussian beam with circularly polarization is launched to two q-plate cells and a SLM, two different-order classical OAM modes with orthogonally-circular polarization are generated. The amplitude and phase of two OAM modes can be adjusted by changing the retardations of the q-plate cells and the SLM [52], which can go through all the points on the hybrid PS, as shown in the third row of Figure 6.

For these kinds of generation method, the greatest advantage is high flexibility due to the phase distributions and retardations can be arbitrarily electrically controlled. However, the common shortcoming is limited response speed.

For the second type, the tunable OAM modes can be generated by interfering two Laguerre–Gaussian modes with same or different topological charges, as shown in Figure 7. The modulation of amplitude is realized by rotating the optical elements [49,50], where "*a*" and "*b*" are the amplitudes of x and y polarizations. The modulation of phase between two Laguerre–Gaussian modes can be realized by rotating the optical elements [48] or changing the phase distribution of SLM/DM [49]. Because the phase difference between the two split and recombined beams determines the properties of the generated vector beams, this kind of method employing interferometry may be vulnerable to environmental noise like vibrations or air circulation. Another weakness is the slow rotating speed for the optical elements.

The third type of generating the beam with tunable OAM in the free-space system is achieved by continuously changing the state of input polarization [51,55–59], as shown in Figure 8. A Gaussian beam with arbitrary polarization state can be generated by using a polarizer followed by an arbitrarily oriented QWP. The arbitrary polarization state can be expressed as the $A\vec{x} + B\vec{y}$ in the basis of x and y polarizations or $A\sigma^{+} + B\sigma^{-}$ in the basis of right- and left-hand circular polarizations, where the symbols "A" and "B" are complex amplitudes.

When a Gaussian mode with arbitrary polarization is injected to the combination of a SLM, a SPP/DM and a QWP, the SLM is used to generate the helical phase distributions of x and y polarizations. The SPP/DM is for compensating superfluous phase factor, and the QWP can convert the orthogonal linear polarizations to orthogonal circular polarizations. As shown in the top branch of Figure 8, when $\varphi = l\theta$, the modes on the higher-order PS can be generated by adjusting the input polarization [50]. When $\varphi = m\theta$, where $m \neq l$, the system can generate the modes on the hybrid-order PS. If the input beam is launched to a q-plate, the output mode will be located on higher-order PS according to Equation (20), as shown in the bottom branch of Figure 8 [53–57]. In addition, by adding a phase factor $\exp(i(l + n)\theta/2)$ into the bottom branch of Figure 8, the tunable OAM on the hybrid-order PS can also be generated by adjusting the input polarization [52,55]. The extra phase can be achieved by numerous methods, such as spiral phase plates, SLM, diffractive elements and fork gratings.

This kind of generation methods involve fewer components. The speed of adjusting polarization is very fast, which leads to a rapid conversion between modes on the PS.

We conclude the spatial generation methods, as shown in Table 1. The table lists the adjustable variations, devices and the types of tunable OAM modes.

Table 1. The free-space generation methods.

Adjustable Variation	Reference	Device	The Type of Tunable OAM
Phase distribution	[17,42–45]	SLM	Orbital PS
	[46,47]	SLM + QWP	Three PSs
	[51]	DMs	Orbital PS
	[52]	Q-plate	Hybrid-order PS
Interference	[48,49]	SLM	Higher- or Hybrid-order PS
Input polarization	[50]	SLM + QWP	Higher-order PS
	[52–55]	Q plate + SPP	Higher- or Hybrid-order PS
	[56]	Metasurface	Higher-order PS

4.2. *The Fiber-Based Generation of Tunable OAM*

For the fiber-based devices, the methods for tunable OAM mode generation can be classified into three types according to the adjusting schemes. Figure 9 simply shows schematic diagram usually used to generate the tunable OAM mode based on the fiber. The components marked by green arrows are the adjusted ones in the experiment.

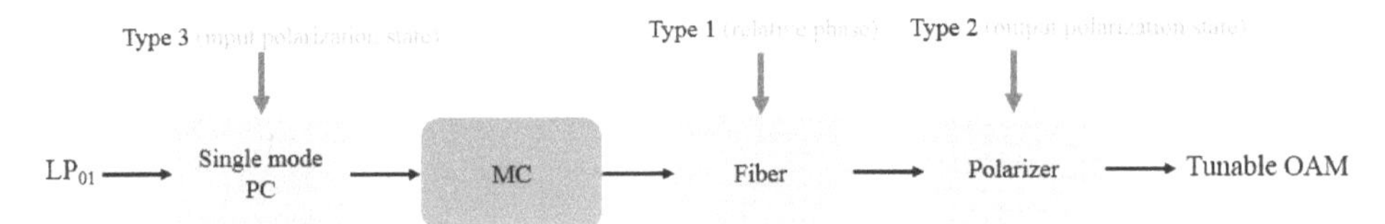

Figure 9. A typical experimental setup based on fiber to generate tunable OAM modes. PC: polarization controller, MC: mode converter.

For the first type, the tunable OAM modes can be generated by the superposition of two orthogonal LP or HE (EH) modes with tunable relative phase difference between LP (HE) modes passing through a polarizer with a fixed direction. The flow chart of the mode generation is shown in Figure 10, and the yellow parts are the variations that need to be adjusted. The tunable OAM on the orbital PS can be generated by using the first four formulas, and the tunable OAM on the higher-order PS can be generated by the last one. Firstly, the LP_{01} mode can be converted to LP_{lm} mode by many kinds of mode converters, for example, photonic lanterns [57], mode selective couplers [58] and gratings [59]. The LP_{lm} mode can be thought of as the combining result of LP_{lma} and LP_{lmb} without relative phase. The HE and EH modes can be directly generated by gratings with appropriate period [60]. When the converted two LP or HE (EH) modes pass through a length L of few-mode fiber (FMF), the relative phase between the two modes at the output of the FMF can always be written as $\Delta\delta = \frac{2\pi L \Delta neff}{\lambda}$. The Δneff represents the RI difference between two modes and λ represents the operating wavelength. Thus, in order to achieve flexible control of the relative phase between the two modes in the fiber, changing the operating wavelength λ [61,62] and the refractive index difference Δneff are two commonly used methods. So far, ways of controlling Δneff mainly depend on adjusting the pressure loaded on the fiber and rotating the paddles of few-mode polarization controller [63–65]. In addition, the LP_{11a} and LP_{11b} can be generated by respectively injecting two LP_{01} modes with slight horizontal and vertical displacement from the fiber axis and the relative phase can be controlled by using a piezo-driven delay stage [26,27,66].

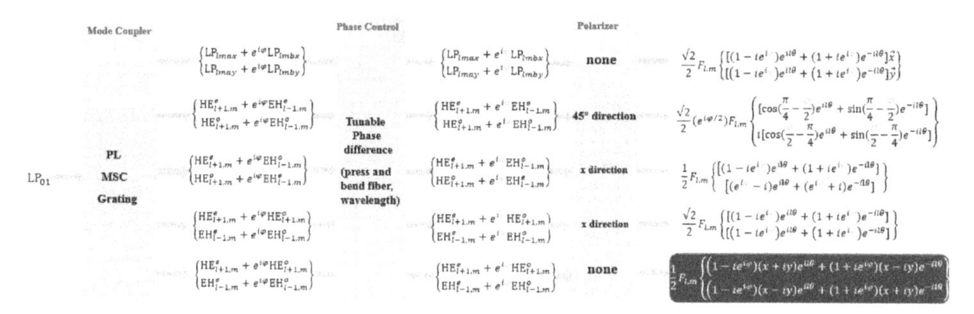

Figure 10. The flow chart of tunable OAM generation by adjusting the relative phase between two fiber modes.

The second type, the tunable OAM modes can be achieved by filtering the mixing modes which are produced by the combination of different vector modes or two spatially orthogonal LP modes owning orthogonal polarization directions with a $\pm\pi/2$ phase shift. The phase shift can be obtained in the same ways as mentioned above. Then, the continually tunable OAM can be achieved by adjusting direction of the polarizer at the output of the FMF [62,63,66,67]. The specific process is shown in Figure 11 and the "p" is the angle between the direction of the polarizer and the positive direction of the x-axis.

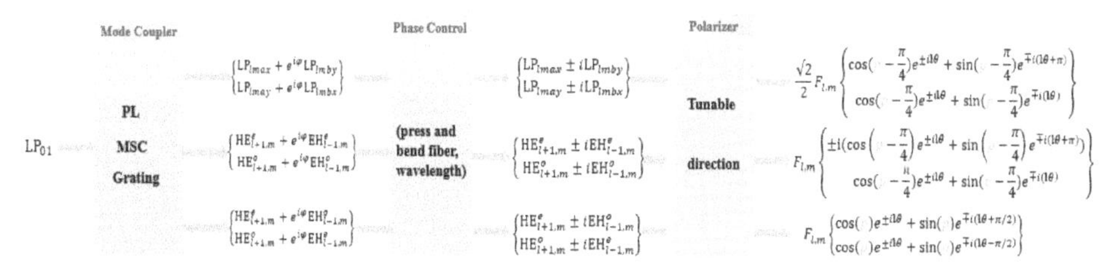

Figure 11. The flow chart of tunable OAM generation by adjusting the polarization direction.

For the third type, a method is reported to generate the beam with tunable OAM in the fiber by continuously changing the angle of linear polarization state of the input light [67], as described in Figure 12. The setup is composed of three parts, including a mode converter, a few-mode fiber that is mounted as coils in a paddle of a fiber polarization controller (PMC) and a polarizer. Considering about four linear polarization (LP) mode bases in the fiber, we deduce the transmission matrix of the first-order modes in PMC. Then, one polarization is filtered out through a polarizer. It is well known that the FMF is wound around the circumference of the PMC's paddle, and stress will induce the refractive RI difference between four orthogonal LP mode bases. If the relative phases between LP_{11ax} and LP_{11bx}, LP_{11ay} and LP_{11by} are $\pi/2$ and $-\pi/2$, the average OAM value of mode will smoothly vary from -1 to 1 with the input light polarization angle changing from 0 to π. The polarization angle (α) can be adjusted by electrical polarization controller. δ in the Figure 10 represents the relative phase between LP_{11bx} and LP_{11by}, which decides the orientation of longitude on the orbital PS.

At the end of the paper, we draw conclusions about the fiber-based generation methods, as shown in Table 2. The table lists the combination modes, adjusting variations, adjusting methods and the types of tunable OAM mode.

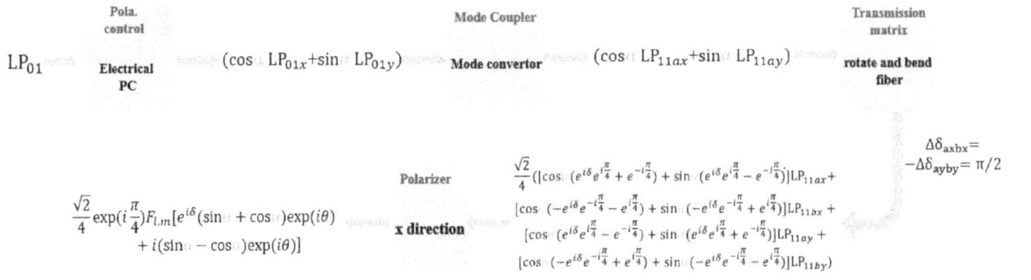

Pola. control Electrical PC
Mode Coupler Mode convertor
Transmission matrix rotate and bend fiber
LP_{01} → $(\cos LP_{01x}+\sin LP_{01y})$ → $(\cos LP_{11ax}+\sin LP_{11ay})$
$\Delta\delta_{axbx} = -\Delta\delta_{ayby} = \pi/2$
Polarizer x direction
$\frac{\sqrt{2}}{4}\exp(i\frac{\pi}{4})F_{l,m}[e^{i\delta}(\sin + \cos)\exp(i\theta) + i(\sin - \cos)\exp(i\theta)]$

Figure 12. The flow chart of tunable OAM generation by adjusting the input polarization.

Table 2. The fiber-based generation methods.

Adjustable Variation	Reference	Combination Modes	Adjusting Method	The Type of Tunable OAM
Relative Phase (φ)	[63]	$LP_{11ax(y)}$ and $LP_{11bx(y)}$	Stress the fiber by a pair of flat slabs	Orbital PS
	[61]	$LP_{11ax(y)}$ and $LP_{11bx(y)}$	Operating wavelength λ in the PMF	Orbital PS
	[26,27]	$LP_{11ax(y)}$ and $LP_{11bx(y)}$	Piezo-driven delay stage	Orbital PS
	[62]	$HE^e_{2,1}$ and $TE_{0,m}$ $HE^o_{2,1}$ and $TM_{0,m}$ $HE^e_{2,1}$ and $TM_{0,m}$ $HE^o_{2,1}$ and $TE_{0,m}$	Wavelength λ in the ring-core fiber	Orbital PS
	[64]	$HE^e_{2,1}$, HE^o_{21}	Bend and twist RCF by paddle-type polarization controller	Orbital and higher-order PS
Polarization direction	[60,65]	$HE^e_{2,1}$ and $TM_{0,m}$ $TE_{0,m}$ and HE^o_{21}	Rotate polarizer	Orbital PS
	[62,66]	LP_{11ax} and LP_{11by} (LP_{11ax} and LP_{11by})	Rotate polarizer	Orbital PS
	[62]	$HE^e_{2,1}$, HE^o_{21} $TE_{0,m}$, $TM_{0,...}$	Rotate polarizer	Orbital PS
Input polarization	[67]	LP_{11a}	Adjust single mode PC	Orbital PS

5. Discussion and Perspective

Arbitrarily tunable OAM has excited a great diversity of interest, because of a variety of emerging applications, but its creation still remains a tremendous challenge. We review the concepts of general OAM, which extends the OAM carried by the scalar vortex modes (classical OAM mode and the modes on the orbital PS) and the OAM carried by the azimuthally varying polarized vector modes (the modes on the higher-order PS and hybrid-order PS).

In summary, due to unique characteristics, tunable OAM beams have been the subject of much interest for a variety of fundamental research studies and modern applications. There are mainly two types of methods to generate those tunable OAM beams, free-space and fiber generating methods. Each method has its own advantages and disadvantages. Free-space generating methods have advantages in terms of flexible design and easy manipulation, but active optical spatial phase modulators are expensive and may introduce additional electronic noise. Meanwhile, the volumes of spatial devices are usually large. Compared with free-space generation methods, the fiber-based generation methods have the advantages of the miniaturization and low insertion loss. However, the challenge is robustness because they are basically the combination of fibers and mode converters. The fiber is vulnerable to external influences and some methods involve alignment operation. It must be mentioned that choosing which method to generate tunable OAM modes depends more on its application scenarios. For example, for a transfer of ultrashort pulses, free-space solutions have essential advantages because of avoiding frequency chirping and pulse lengthening. For medical endoscopy, the fiber-based generation method is obviously a better way. Despite this, along with the efforts of researchers all around the world, we may see an increasing number of applications based on tunable OAM beams in the future.

Author Contributions: This paper was mainly wrote by L.F. and Y.L. and the idea originates from Y.L., S.W. and W.L. provided constructive improvements and feedback. All authors contributed to the writing and editing of the manuscript.

Funding: This paper is partly supported by National Natural Science Foundation of China (NSFC) (61875019, 61675034, 61875020, 61571067); The Fund of State Key Laboratory of IPOC (BUPT); The Fundamental Research Funds for the Central Universities.

Conflicts of Interest: The authors declare no conflict of interest.

References

1. Allen, L.; Beijersbergen, M.W.; Spreeuw, R.J.C.; Woerdman, J.P. Orbital angular-momentum of light and the transformation of Laguerre–Gaussian laser modes. *Phys. Rev. A* **1992**, *45*, 8185–8189. [CrossRef] [PubMed]
2. Padgett, M.; Bowman, R. Tweezers with a twist. *Nat. Photonics* **2011**, *5*, 343–348. [CrossRef]
3. Paterson, L. Controlled Rotation of Optically Trapped Microscopic Particles. *Science* **2001**, *292*, 912–914. [CrossRef] [PubMed]
4. Simpson, N.B.; Dholakia, K.; Allen, L.; Padgett, M.J. Mechanical equivalence of spin and orbital angular momentum of light: An optical spanner. *Opt. Lett.* **1997**, *22*, 52–54. [CrossRef] [PubMed]
5. Padgett, M.; Allen, L. Optical tweezers and spanners. *Phys. World* **1997**, *10*, 35–38. [CrossRef]
6. Molina-Terriza, G.; Torres, J.P.; Torner, L. Management of the angular momentum of light: Preparation of photons in multidimensional vector states of angular momentum. *Phys. Rev. Lett.* **2002**, *88*, 013601. [CrossRef] [PubMed]
7. Krenn, M.; Malik, M.; Erhard, M.; Zeilinger, A. Orbital angular momentum of photons and the entanglement of Laguerre–Gaussian modes. *Philos. Trans. R. Soc. A-Math. Phys. Eng. Sci.* **2017**, *375*, 20150442. [CrossRef]
8. Jack, B.; Yao, A.M.; Leach, J.; Romero, J.; Franke-Arnold, S.; Ireland, D.G.; Barnett, S.M.; Padgett, M.J. Entanglement of arbitrary superpositions of modes within two-dimensional orbital angular momentum state spaces. *Phys. Rev. A* **2010**, *81*, 043844. [CrossRef]
9. Mair, A.; Vaziri, A.; Weihs, G.; Zeilinger, A. Entanglement of Orbital Angular Momentum States of Photons. *Nature* **2001**, *412*, 313–316. [CrossRef]
10. Vaziri, A.; Weihs, G.; Zeilinger, A. Experimental Two-Photon, Three-Dimensional Entanglement for Quantum Communication. *Phys. Rev. Lett.* **2002**, *89*, 240401. [CrossRef]
11. Fürhapter, S.; Jesacher, A.; Bernet, S.; Ritsch-Marte, M. Spiral phase contrast imaging in microscopy. *Opt. Express* **2005**, *13*, 689–694. [CrossRef] [PubMed]
12. Yan, L.; Gregg, P.; Karimi, E.; Rubano, A.; Marrucci, L.; Boyd, R.; Ramachandran, S. Q-plate enabled spectrally diverse orbital-angular-momentum conversion for stimulated emission depletion microscopy. *Optica* **2015**, *2*, 900–903. [CrossRef]
13. Bernet, S.; Jesacher, A.; Severin, F.; Maurer, C.; Ritsch-Marte, M. Quantitative imaging of complex samples by spiral phase contrast microscopy. *Opt. Express* **2006**, *14*, 3792–3805. [CrossRef] [PubMed]
14. Padgett, M.J.; Allen, L. Light with a twist in its tail. *Contemp. Phys.* **2000**, *41*, 275–285. [CrossRef]
15. Wang, J. Twisted optical communications using orbital angular momentum. *Sci. China Phys. Mech. Astron.* **2019**, *62*, 034201. [CrossRef]
16. Elias, N.M. Photon orbital angular momentum in astronomy. *Astron. Astrophys.* **2008**, *492*, 883–922. [CrossRef]
17. Schmitz, C.H.J.; Uhrig, K.; Spatz, J.P.; Curtis, J.P. Tuning the orbital angular momentum in optical vortex beams. *Opt. Express* **2006**, *14*, 6604–6612. [CrossRef]
18. Gecevicius, M.; Drevinskas, R.; Beresna, M.; Kazansky, P.G. Single beam optical vortex tweezers with tunable orbital angular momentum. *Appl. Phys. Lett.* **2014**, *104*, 231110. [CrossRef]
19. Yao, A.M.; Padgett, M.J. Orbital angular momentum: Origins, behavior and applications. *Adv. Opt. Photonics* **2011**, *3*, 161–204. [CrossRef]
20. Gibson, G.; Courtial, J.; Padgett, M.J.; Vasnetsov, M.; Pas'ko, V.; Barnett, S.M.; Franke-Arnold, S. Free-space information transfer using light beams carrying orbital angular momentum. *Opt. Express* **2004**, *12*, 5448–5456. [CrossRef]
21. Heckenberg, N.R.; Mcduff, R.; Smith, C.P.; White, A.G. Generation of optical phase singularities by computer-generated holograms. *Opt. Lett.* **1992**, *17*, 221–223. [CrossRef] [PubMed]

22. Beijersbergen, M.W.; Coerwinkel, R.P.C.; Kristensen, M.; Woerdman, J.P. Helical-wavefront laser beams produced with a spiral phase plate. *Opt. Commun.* **1994**, *112*, 321–327. [CrossRef]
23. Beijersbergen, W.M.; Allen, V.D.; Veen, E.L.; Woerdman, O.H. Astigmatic laser mode converters and transfer of orbital angular momentum. *Opt. Commun.* **1993**, *96*, 123–132. [CrossRef]
24. Marrucci, L.; Karimi, E.; Slussarenko, S.; Piccirillo, B.; Santamato, E.; Nagali, E.; Sciarrino, F. Spin-to-orbital conversion of the angular momentum of light and its classical and quantum applications. *J. Opt.* **2011**, *13*, 064001. [CrossRef]
25. Ramachandran, S. Optical Vortices in Fiber. *Nanophotonics* **2013**, *2*, 455–474. [CrossRef]
26. Niederriter, R.D.; Siemens, M.E.; Gopinath, J.T. Continuously tunable orbital angular momentum generation using a polarization-maintaining fiber. *Opt. Lett.* **2016**, *41*, 3213–3216. [CrossRef]
27. Niederriter, R.D.; Siemens, M.E.; Gopinath, J.T. Simultaneous control of orbital angular momentum and beam profile in two-mode polarization-maintaining fiber. *Opt. Lett.* **2016**, *41*, 5736–5739. [CrossRef]
28. Padgett, M.J.; Courtial, J. Poincare-sphere equivalent for light beams containing orbital angular momentum. *Opt. Lett.* **1999**, *24*, 430–432. [CrossRef]
29. Kumar, A.; Ajoy, G. Poincaré Sphere Representation of Polarized Light. In *Polarization of Light with Applications in Optical Fibers*; SPIE: Bellingham WA, USA, 2011; Volume TT90, pp. 121–134.
30. Zhan, Q. Cylindrical vector beams: From mathematical concepts to, applications. *J. Syst. Sci. Complex.* **2009**, *27*, 899–910. [CrossRef]
31. Deng, D.; Guo, Q. Analytical vectorial structure of radially polarized light beams. *Opt. Lett.* **2007**, *32*, 2711–2713. [CrossRef]
32. Kozawa, Y.; Sato, S. Optical trapping of micrometer-sized dielectric particles by cylindrical vector beams. *Opt. Express* **2010**, *18*, 10828–10833. [CrossRef] [PubMed]
33. Huang, L.; Guo, H.; Li, J.; Ling, L.; Feng, B.; Li, Z. Optical trapping of gold nanoparticles by cylindrical vector beam. *Opt. Lett.* **2012**, *37*, 1694–1696. [CrossRef] [PubMed]
34. Nesterov, A.V.; Niziev, V.G. Laser beams with axially symmetric polarization. *J. Appl. Phys.* **2000**, *33*, 1817–1822. [CrossRef]
35. Zhou, J.; Yang, L.; Wang, S.; Zhou, J. Minimized spot of annular radially polarized focusing beam. *Opt. Lett.* **2013**, *38*, 1331–1333.
36. Milione, G.; Sztul, H.I.; Nolan, D.A.; Alfano, R.R. Higher-Order Poincaré Sphere, Stokes Parameters, and the Angular Momentum of Light. *Phys. Rev. Lett.* **2011**, *107*, 053601. [CrossRef]
37. Holleczek, A.; Aiello, A.; Gabriel, C.; Marquardt, C.; Leuchs, G. Classical and quantum properties of cylindrically polarized states of light. *Opt. Express* **2011**, *19*, 9714–9736. [CrossRef] [PubMed]
38. Roxworthy, B.J.; Toussaint, K.C.J. Optical trapping efficiencies from n-phase cylindrical vector beams. *Proc. SPIE* **2011**, *7950*, 2362–2375.
39. Zhao, Z.; Wang, J.; Li, S.; Willner, A.E. Metamaterials-based broadband generation of orbital angular momentum carrying vector beams. *Opt. Lett.* **2013**, *38*, 932–934. [CrossRef]
40. Qiu, C.; Palima, D.; Novitsky, A.; Gao, D.; Ding, W.; Zhukovsky, S.V.; Gluckstad, J. Engineering light-matter interaction for emerging optical manipulation applications. *Nanophotonics* **2014**, *3*, 181–201. [CrossRef]
41. Yi, X.; Liu, Y.; Ling, X.; Zhou, X.; Ke, Y.; Luo, H.; Wen, S.; Fan, D. Hybrid-order Poincare sphere. *Phys. Rev. A* **2015**, *91*, 023801. [CrossRef]
42. Schmitz, C.H.J.; Spatz, J.P.; Curtis, J.E. High-precision steering of holographic optical tweezers. *Opt. Express* **2005**, *13*, 8678–8685. [CrossRef] [PubMed]
43. Grier, D.G. A revolution in optical manipulation. *Nature* **2003**, *424*, 810–816. [CrossRef] [PubMed]
44. Cizmar, T.; Dalgarno, H.I.C.; Ashok, P.C.; Gunn-Moore, F.J.; Dholakia, K. Interference-free superposition of nonzero order light modes: Functionalized optical landscapes. *Appl. Phys. Lett.* **2011**, *98*, 081114. [CrossRef]
45. Curtis, E.; Grier, G. Modulated optical vortices. *Opt. Lett.* **2003**, *28*, 872–874. [CrossRef] [PubMed]
46. Han, W.; Yang, Y.; Cheng, W.; Zhan, Q. Vectorial optical field generator for the creation of arbitrarily complex fields. *Opt. Express* **2013**, *21*, 20692–20706. [CrossRef] [PubMed]
47. Moreno, I.; Davis, J.A.; Hernandez, T.M.; Cottrell, D.M.; Sand, D. Complete polarization control of light from a liquid crystal spatial light modulator. *Opt. Express* **2012**, *20*, 364–376. [CrossRef] [PubMed]
48. Chen, S.; Zhou, X.; Liu, Y.; Ling, X.; Luo, H. Generation of arbitrary cylindrical vector beams on the higher order Poincare sphere. *Opt. Lett.* **2014**, *39*, 5274–5276. [CrossRef]

49. Maurer, C.; Jesacher, A.; Fürhapter, S.; Bernet, S.; Ritsch-Marte, M. Tailoring of arbitrary optical vector beam. *New J. Phys.* **2007**, *9*, 78. [CrossRef]
50. Tripathi, S.; Toussaint, K.C. Versatile generation of optical vector fields and vector beams using a non-interferometric approach. *Opt. Express* **2012**, *20*, 10788–10795. [CrossRef] [PubMed]
51. Chensheng, W.; Jonathan, K.; Rzasa, J.R.; Paulson, D.A.; Davis, C.C. Phase and amplitude beam shaping with two deformable mirrors implementing input plane and Fourier plane phase modifications. *Appl. Opt.* **2018**, *57*, 2337–2345.
52. Lou, S.; Zhou, Y.; Yuan, Y.; Lin, T.; Fan, F.; Wang, X.; Huang, H.; Wen, S. Generation of arbitrary vector vortex beams on hybrid-order Poincaré sphere based on liquid crystal device. *Opt. Express* **2019**, *27*, 8596–8604. [CrossRef] [PubMed]
53. Gregg, P.; Mirhosseini, M.; Rubano, A.; Marrucci, L.; Karimi, E.; Boyd, R.W.; Ramachandran, S. Q-plates as higher order polarization controllers for orbital angular momentum modes of fiber. *Opt. Lett.* **2015**, *40*, 1729–1732. [CrossRef] [PubMed]
54. Darryl, N.; Filippus, S.R.; Dudley, A.; Litvin1, I.; Piccirillo, B.; Marucci, L.; Forbes, A. Controlled generation of higher-order Poincare sphere beams from a laser. *Nat. Photonics* **2016**, *10*, 327–332.
55. Liu, Z.; Liu, Y.; Ke, Y.; Liu, Y.; Shu, W.; Luo, H.; Wen, S. Generation of arbitrary vector vortex beams on hybrid-order Poincaré sphere. *Photonics Res.* **2017**, *1*, 19–25. [CrossRef]
56. Liu, Y.; Ling, X.; Yi, X.; Zhou, X.; Luo, H.; Wen, S. Realization of polarization evolution on higher-order Poincare sphere with metasurface. *Appl. Phys. Lett.* **2014**, *104*, 1–4. [CrossRef]
57. Birks, T.A.; Gris-Sánchez, I.; Yerolatsitis, S.; Leon-Saval, S.G.; Thomson, R.R. The photonic lantern. *Adv. Opt. Photonics* **2015**, *7*, 107–167. [CrossRef]
58. Park, K.J.; Song, K.Y.; Kim, Y.K.; Lee, J.H.; Kim, B.Y. Broadband mode division multiplexer using all-fiber mode selective couplers. *Opt. Express* **2016**, *24*, 3543–3549. [CrossRef]
59. Zhao, Y.; Chen, H.; Fontaine, N.K.; Li, J.; Ryf, R.; Liu, Y. Broadband and low-loss mode scramblers using CO_2-laser inscribed long-period gratings. *Opt. Lett.* **2018**, *43*, 2868–2871. [CrossRef]
60. Jiang, Y.; Ren, G.; Lian, Y.; Zhu, B.; Jin, W.; Jian, S. Tunable orbital angular momentum generation in optical fibers. *Opt. Lett.* **2016**, *41*, 3535–3538. [CrossRef]
61. Zeng, X.; Li, Y.; Mo, Q.; Li, W.; Tian, Y.; Liu, Z.; Wu, J. Experimental Investigation of LP11 Mode to OAM Conversion in Few Mode-Polarization Maintaining Fiber and the Usage for All Fiber OAM Generator. *IEEE Photonics J.* **2017**, *8*, 1–7.
62. Jiang, Y.; Ren, G.; Shen, Y.; Xu, Y.; Jin, W.; Wu, Y.; Jian, W.; Jian, S. Two-dimensional tunable orbital angular momentum generation using a vortex fiber. *Opt. Lett.* **2017**, *42*, 5014–5017. [CrossRef] [PubMed]
63. Li, S.; Mo, Q.; Hu, X.; Du, C.; Wang, J. Controllable all-fiber orbital angular momentum mode converter. *Opt. Lett.* **2015**, *40*, 4376–4379. [CrossRef] [PubMed]
64. Bozinovic, N.; Golowich, S.; Kristensen, P.; Ramachandran, S. Control of orbital angular momentum of light with optical fibers. *Opt. Lett.* **2012**, *37*, 2451–2453. [CrossRef] [PubMed]
65. Yao, S.; Ren, G.; Shen, Y.; Jiang, Y.; Zhu, B.; Jian, S. Tunable Orbital Angular Momentum Generation Using All-Fiber Fused Coupler. *IEEE Photonics Tech. Lett.* **2017**, *30*, 99–102. [CrossRef]
66. Jiang, Y.; Ren, G.; Li, H.; Tang, M.; Jin, W.; Jian, W.; Jian, S. Tunable Orbital Angular Momentum Generation Based on Two Orthogonal LP Modes in Optical Fibers. *IEEE Photonics Tech. Lett.* **2017**, *29*, 901–904. [CrossRef]
67. Sihan, W.; Yan, L.; Lipeng, F.; Li, W.; Qiu, J.; Zuo, Y.; Hong, X.; Yu, H.; Chen, R.; Giles, I.P.; et al. Continuously tunable orbital angular momentum generation controlled by input linear polarization. *Opt. Lett.* **2018**, *43*, 2130–2133.

Article

Nonlinear Metasurface for Structured Light with Tunable Orbital Angular Momentum

Yun Xu [1,2], Jingbo Sun [1], Jesse Frantz [3], Mikhail I. Shalaev [1], Wiktor Walasik [1], Apra Pandey [4], Jason D. Myers [3], Robel Y. Bekele [5], Alexander Tsukernik [6], Jasbinder S. Sanghera [3] and Natalia M. Litchinitser [1,*]

1. Department of Electrical and Computer Engineering, Duke University, Durham, NC 27708, USA; yun.xu746@duke.edu (Y.X.); jingbo.sun482@duke.edu (J.S.); mikhail.shalaev@duke.edu (M.I.S.); wiktor.walasik@duke.edu (W.W.)
2. Department of Electrical Engineering, The State University of New York, University at Buffalo, Buffalo, NY 14260, USA
3. US Naval Research Laboratory, Washington, DC 20375, USA; jesse.frantz@nrl.navy.mil (J.F.); jason.myers@nrl.navy.mil (J.D.M.); jas.sanghera@nrl.navy.mil (J.S.S.)
4. CST of America, LLC, Santa Clara, CA 95054, USA; pandeyapra@gmail.com
5. University Research Foundation, Greenbelt, MD 20770, USA; robel.bekele.ctr@nrl.navy.mil
6. Toronto Nanofabrication Centre, University of Toronto, Toronto, ON M5S 3G4, Canada; alex.tsukernik@utoronto.ca

* Correspondence: natalia.litchinitser@duke.edu

Received: 7 February 2019; Accepted: 2 March 2019; Published: 6 March 2019

Abstract: Orbital angular momentum (OAM) beams may create a new paradigm for the future classical and quantum communication systems. A majority of existing OAM beam converters are bulky, slow, and cannot withstand high powers. Here, we design and experimentally demonstrate an ultra-fast, compact chalcogenide-based all-dielectric metasurface beam converter which has the ability to transform a Hermite–Gaussian (HG) beam into a beam carrying an OAM at near infrared wavelength. Depending on the input beam intensity, the topological charge carried by the output OAM beam can be switched between positive and negative. The device provides high transmission efficiency and is fabricated by a standard electron beam lithography. Arsenic trisulfide (As_2S_3) chalcogenide glass (ChG) offers ultra-fast and large third-order nonlinearity as well as a low two-photon absorption coefficient in the near infrared spectral range.

Keywords: nonlinear optics; metasurfaces; structured light

1. Introduction

Structured light and, in particular, beams carrying orbital angular momentum (OAM) have been shown to enable and expand a plethora of photonic applications from optical trapping and manipulation, to astronomy and light filamentation [1]. Moreover, the OAM of light can be used as an alternate degree of freedom for expanding the capacity of communication channels [2–6]. Many of these systems require the development of dynamically reconfigurable and high-power OAM beams. Usually, the OAM beams are generated using bulk optical devices such as spiral phase plates (SPPs) and spatial light modulators (SLMs) [7]. However, an SPP is only suitable to generate an OAM beam with fixed topological charge for a designed wavelength. SLMs, dynamically controlled by computers, are able to generate tunable OAM beams with high intensities [8] but are limited by their resolution, bulky dimensions, and the switching speed. For nonlinear applications requiring dynamically changing topological charge of light together with high intensities [9], a tunable OAM metasurface-beam converter may enable new opportunities. To date, the realization of tunable OAM beam converters that can be used in high-power applications with ultrafast switching speed and can be incorporated in micro-scale systems remains a grand challenge.

Here, we propose and demonstrate a nonlinear metasurface based on chalcogenide glass (ChG) to generate reconfigurable topological charge, depending on the input intensity, and able to generate OAM beam with an input intensity higher than 4 GW/cm^2. Photonic metasurfaces attracted significant attention owing to their compact size, flat topology, and compatibility with existing integrated-optics fabrication methods [10–31]. Recently, we have demonstrated the first step toward the realization of input-intensity-dependent optical metasurfaces capable of converting a beam with no OAM into an OAM-carrying beam in the near infrared range [32]. Here, we describe a nonlinear metasurface with the capability of switching between two opposite topological charges of OAM beams depending the intensity of the input beam.

The proposed metasurface beam converter consisted of an array of nano-blocks made of As_2S_3 ChG with high refractive index, low absorption and a very good nonlinear figure of merit in both the near-infrared and the mid-wave infrared spectral bands [33–35]. In addition, the fabrication procedure was a single-step electron-beam lithography since As_2S_3 is sensitive to electron beam exposure [36].

2. Results

2.1. Beam-Converter Metasurface Design

The operation principle of our reconfigurable metasurface is illustrated in Figure 1. The metasurface transformed an input Hermite–Gaussian (HG) beam into an OAM beam with counter-clockwise or clockwise wavefront, depending on the intensity of the incoming light. In the linear regime—for low input intensity—the phase acquired in the even quadrants (II and IV) was 90° larger than in the odd quadrants (I and III). Consequently, the input HG beam was transmitted through the metasurface and the resulting phase distribution directly after the metasurface is given by $\phi = (N-1)90°$, where N enumerated the quadrants. Although in the near-field the phase of the transmitted beam changes in a stepwise manner, in the far-field, the wavefront became helical with a counter-clockwise direction of rotation corresponding to a positive OAM, as shown in Figure 1a. Furthermore, the metasurface was designed such that for a high-intensity input beam, the phase introduced in the even quadrants was increased by 180° respect to the low-intensity regime. Figure 1b shows that in this case, the phase distribution at the output increased in the clockwise direction. In contrast with the low-intensity regime shown in Figure 1a, for high input intensity, the output beam acquired a negative OAM. The reconfigurability of the output beam was enabled by the design of the metasurface utilizing Mie-resonances in conjunction with highly nonlinear ChGs, as described below in detail.

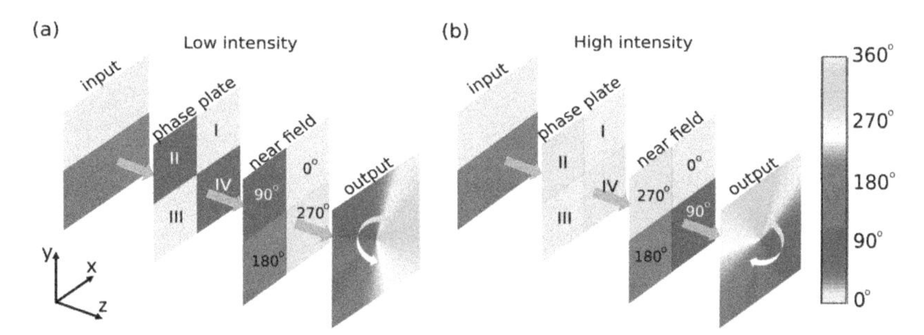

Figure 1. Working principle of a nonlinear metasurface with reconfigurable output beam. The color maps show the phase of the input and output beams, and the phase shifts introduced by the metasurface. (**a**) At low intensity, upon transmission through the metasurface, the input Hermite–Gaussian (HG) beam acquires a non-uniform phase distribution leading to generation of a beam carrying a positive orbital angular momentum (OAM). (**b**) At high intensity, the phase in even quadrants changes by 180°; the output beam possesses a negative OAM, as opposed to the low-intensity case.

In this section, we will demonstrate the design principle of the metasurface that converts HG beams into OAM beams. In Section 2.2, we compare the results of the experimental measurements with simulations for the mode converter operating in the linear regime, described in Section 2.1. Finally, in Section 2.3, we present a design of a nonlinear meatsuface capable of switching between the positive and negative OAM.

The functionality described in Figure 1 is enabled by a ChG-based metasurface illustrated in Figure 2a, which is an array of ChG nano-blocks fabricated on a substrate. The size of the array was assumed to be infinite in the x- and y-directions, and the incident light was a plane wave propagating along the z-axis with electric and magnetic fields polarized along the x- and y-directions, respectively. For the wavelength of interest around 1550 nm, the numerical simulations were performed by CST Microwave Studio with the following geometric parameter: the height of the blocks $h = 400$ nm, the lattice constant $a = 930$ nm, and the side-length of the blocks $l = 700$ nm. The refractive index of ChG was $n_0 = 2.4$ measured using spectroscopic ellipsometry and the refractive index of the glass substrate $n_g = 1.5$. Two dips located at wavelengths 1464 nm and 1510 nm corresponding to resonant interaction with the metasurfaces were indicated on the transmittance spectrum shown in Figure 2b. The near-zero transmittance resulted from a near-unity reflectance, as the materials were assumed to be lossless in the simulations. The distributions of the electric and magnetic fields in the unit cell cross-section are shown in Figure 2c,d. At the wavelength of 1464 nm, the magnetic field formed a vortex around the electric field revealing an electric resonance. At the wavelength of 1510 nm, the vortex-like electric field distribution was a signature of a magnetic resonance. As illustrated in Figure 3, in the vicinity of the central wavelengths of these resonances, the phase of the transmitted light changed rapidly by 180°.

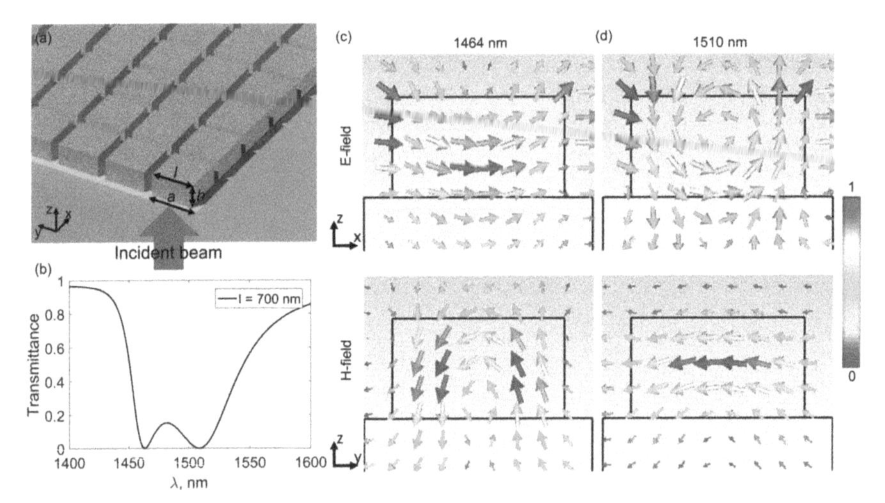

Figure 2. Design of the metasurface. (**a**) Metasurface consisting of a square lattice of square-blocks made of arsenic trisulfide (As_2S_3) chalcogenide glass (ChG) on a glass substrate. The geometric parameters of the metasurface are: height $h = 400$ nm; lattice constant $a = 930$ nm; and the side-length of the blocks is denoted by l. The refractive index of the ChG film is $n_0 = 2.43$. The refractive index of glass is $n_g = 1.5$. (**b**) The transmittance of an array of ChG blocks with $l = 700$ nm. Two resonances are indicated at the wavelengths of 1464 nm and 1510 nm. (**c**) Electric and magnetic fields in the center of the unit cell (cross-section $y = 0$) at the wavelength of 1464 nm. (**d**) Electric and magnetic fields in the center of the unit cell (cross-section $x = 0$) at the wavelength of 1510 nm.

The wavelengths corresponding to the electric and magnetic resonances can be controlled by changing the side-length of the nano-blocks. As the side-length increased, the central wavelengths of both the electric and magnetic resonances increased, and at the same time they become closer to each other. Once the two resonances overlap, a nearly unitary transmission with the phase change spanning 360° can be achieved [25,26,37]. In contrast, when the side-length decreased, the electric and magnetic resonances were shifted to shorter wavelengths and further away from the wavelength of interest. To obtain the desired phase difference for the design shown in Figure 1a, numerical simulations were performed with the side-length of the nano-block swept from $l = 400$ nm to $l = 900$ nm. Figure 3 shows the spectra of the transmittance and the phase of the transmitted light through the metasurface with different side-length for a normally incident light. We chose two different size lengths, $l = 400$ nm and $l = 700$ nm, at the wavelength of 1550 nm, as indicated with the black crosses in Figure 3. The transmittance of both structures was higher than 70% (Figure 3a) and the phase difference between the two structures was 90° (Figure 3b).

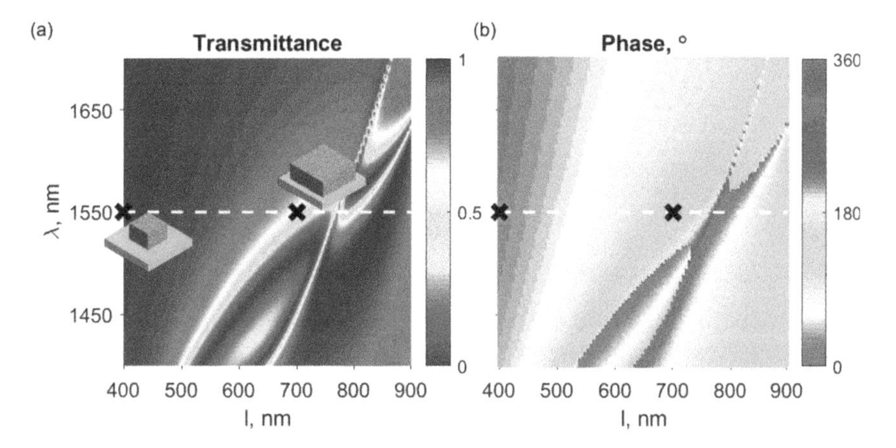

Figure 3. Spectral dependencies of (**a**) the transmittance and (**b**) the phase of the transmitted light as a function of the side-length, revealing the presence of Mie resonances. The dashed line indicates the operation wavelength $\lambda_0 = 1550$ nm. The two black crosses indicate the parameters used for the metasurface beam converter design.

2.2. Simulations and Experiments

In order to demonstrate the proposed OAM beam-converter, a metasurface was fabricated with two sizes of the blocks $l = 400$ nm and $l = 700$ nm. The arrays of nano-blocks were patterned in ChG thin film using electron-beam lithography, as shown in Figure 4a. First, chromium windows were prepared on a glass substrate. Then, a ChG film was deposited on top of the chromium windows with thermal deposition. The linear refractive index of the film was measured using spectroscopic ellipsometry to be $n_0 = 2.43$. Finally, after the exposure to electron beam, a solution of diluted MF-319 was used to develop the sample. The resulting ChG structure with a thickness of 400 nm is shown Figure 4b. The fabricated sample contains four quadrants where diagonal quadrants contain blocks with the same side-length. The total size of the fabricated metasurface was 93 μm × 93 μm.

The simulation of a whole metasurface consisting of four quadrants is performed with CST Microwave Studio. Then we simulate the propagation of the near-field result from CST in free space for 4 mm with the beam propagation method [38]. Figure 4c shows the normalized intensity and Figure 4d illustrates the phase of transmitted light. The spiral-shaped phase in Figure 4d proves that the wavefront of the transmitted beam is helical.

The fabricated metasurface was characterized using the interferometry setup shown in Figure 4e. The beam from a photodiode laser at the wavelength of 1550 nm was split by a beam splitter. The main beam was focused on the sample by a lens and collimated by a second lens placed after the sample. Then the main beam was combined with the reference beam using a beam splitter and the resulting interference patterns were captured by a camera. The experimental measurement of the main beam transmitted through the metasurface is shown in Figure 4f. The intensity profile with a dark singularity at the center suggests that the beam carries an OAM. In order to prove the presence of the helical wavefront, we performed two experiments on the interference of the main beam and the reference Gaussian beam. In the first experiment, after the two beams were combined, the propagating directions and centers of the Gaussian beam and main OAM beam were overlapped. The resulting interference pattern reveals spiral-shaped fringes as shown in Figure 4g. In the second experiment, a small angle was introduced between the two beams, and the interference pattern contained fork-like fringes indicating the presence of the OAM, as shown in Figure 4h.

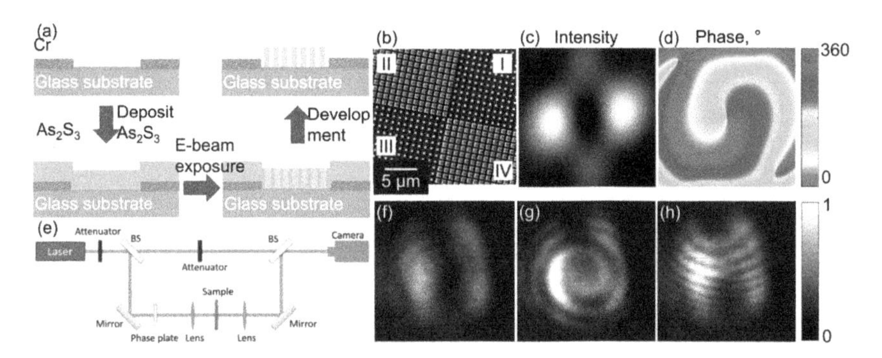

Figure 4. (**a**) Fabrication process for ChG-based all dielectric metasurfaces involves the following steps: fabrication of a chromium window with the size of 150 µm × 150 µm; deposition of an As_2S_3 film; exposure of the As_2S_3 using electron beam; development of the exposed sample. (**b**) Scanning electron microscopy image of the fabricated metasurface with four quadrants. (**c**) Intensity and (**d**) phase of simulation results of an HG beam transmitted through the metasurface. (**e**) Schematic of the Mach–Zehner interferometer used to characterize the fabricated metasurface. (**f–h**) Experimental results obtained with the fabricated metasurface. (**f**) Intensity distribution of the output OAM beam. (**g**,**h**) Interference of the output OAM beam and a reference Gaussian beam showing a spiral-shaped and fork-like intensity distributions (see text for details).

In our previous work, the ChG hole metasurface converts an HG beam into an OAM beam by utilizing the guided resonances in a photonic crystal structure and overlapping two guided resonances to realize the desired phase shift in the even quadrants [32]. In contrast, in this experiment, the phase different between different quadrants of the fabricated metasurface is introduced by a single Mie-resonance and according to Figure 2, the resonance close to the wavelength of 1550 nm is the magnetic resonance.

2.3. *Reconfigurable Beam with OAM*

In this section, we demonstrate that the third-order nonlinearity (Kerr effect) of ChG can be used to achieve the switching of the output beam OAM from positive to negative. The refractive index of the ChG n_{ChG} can be efficiently changed as a function of the input light intensity due to its large nonlinear Kerr response described by $n_{ChG} = n_0 + \Delta n = n_0 + n_2 I$. Here, n_0 is the linear refractive index of the ChG, $n_2 = 7.9 \times 10^{-13}$ cm^2/W is the nonlinear coefficient measured by our in-home Z-scan setup, and Δn is the refractive index change of the ChG corresponding the input beam intensity I. Figure 5 shows the result of the nonlinear studies of ChG nano-blocks with different side-length. The input intensity was varied from 1 kW/cm^2, which is low enough to avoid significant refractive index change, to 4 GW/cm^2. As shown in Figure 5a,b, for l = 700 nm, around the wavelengths of electric and magnetic resonances indicated by the dashed lines, the phase change of the transmitted light is around 180°. However, at these two wavelength, the transmittance is close to zero as the two resonances are spectrally separated. Therefore, this structure is not suitable to design a nonlinear metasurface. As explained above, with the increase of the side-length, the two resonances shift closer to each other, and as they overlap, nearly 100% transmission can be achieved with the phase change covering the entire 360°. For the side-length 760 nm, the two resonances are both located around the wavelength 1550 nm. As seen in Figure 5, the maximum 180° phase change is achieved at the wavelength of 1572 nm. This wavelength is indicated by the white dashed line in Figure 5a–d and it is the operation wavelength of the OAM beam converter. When the input intensity increases, the phase of the transmitted light changes by 180° and the transmittance at the intensity levels of interest remains higher than 60%, as shown in Figure 5c,d. As illustrated by Figure 1, while the phase in the even quadrants changes 180°, the phase of odd quadrants should remain the same. Therefore, to realize the intensity-dependent switching, we build the odd quadrants with side-length 520 nm for which the resonances are far away from the operation wavelength of 1572 nm. As shown in Figure 5f, for low intensity 1 kW/cm^2, the phase introduced by the structure with side-lengths of 760 nm and 520 nm differs by approximately 90°. The phase of the beam transmitted through the even quadrants (built of the cubes with the side-length of $l = 760$ nm) grows with the increase of the input intensity while the phase in the odd quadrants remains constant. When the intensity increases to 4 GW/cm^2, the phase difference between the odd and even quadrants increases to 270°, while the transmittance remains higher than 60%. Therefore, these two structures can be used to realize the proposed nonlinear metasurface enabling the intensity-dependent OAM switching. The maximum refractive index change inside ChG is $\Delta n = 0.12$ and the maximum intensity inside the ChG blocks is 150 GW/cm^2, as found in the simulation results. The damage threshold of ChG with different compositions measured by the femtosecond laser has been studied by Zhang et al. and You et al. at the near-infrared and mid-infrared wavelengths, respectively [39,40]. Due to the short pulse duration (100 fs) and a low repetition rate (1 KHz), the damage threshold of ChGs is much larger than the peak intensity required for our reconfigurable metasurface. Moreover, it has been reported that if As_2S_3 is properly doped with silver, the nonlinear coefficient of the silver-doped ChG film can be up to two orders of magnitude larger than the nonlinear coefficient of the undoped As_2S_3 film [41]. In this case, the required peak intensity will decrease to approximately 1 GW/cm^2, making the proposed device more energy efficient. Besides energy efficiency, another advantage of the silver-doped ChG with larger nonlinear coefficient is that it may result in a much larger refractive index change Δn. In our current design with the pure As_2S_3, $\Delta n \approx 0.1$. Realization of a 180° phase change with the current value of $\Delta n \approx 0.1$ requires the resonances to be very sharp which places a stringent requirements on the fabrication precision to ensure the rapid phase change, which is beyond our ability right now. However, if a larger Δn can be introduced by the silver-doped ChG, the design may tolerate more fabrication imperfection. The deposition and patterning processes of the silver-doped ChG film will be studied in the future to experimentally realize the reconfigurable metasurface which produces output structured light with tunable topological charges.

When the input intensity increased from 1.94 GW/cm^2 to 1.98 GW/cm^2, the phase of the light transmitted through the structure with the side-length $l = 760$ nm jumped by approximately 90°, as shown in Figure 5e. The origin of this phase jump can be understood by looking at Figure 6 where the electric- and magnetic-field distributions in a unit cell for four selected input intensities are plotted. As shown in Figure 6a,b,e,f, the field distributions were very similar when the input intensity increased from 1 kW/cm^2 to 1.94 GW/cm^2 and they possess only a magnetic-resonance revealed as a vortex-like electric field distribution. When the input intensity increased, the resonances became closer to each other. In Figure 6c,d,g,h, the field distributions are very similar when the input intensity increases from 1.98 GW/cm^2 to 4 GW/cm^2 and both electric- and magnetic-resonances are present at the operation wavelength. As the light intensity increases, the electric-resonance shifts closer to the operation wavelength and as it overlaps with the magnetic resonance it leads to an abrupt phase jump by 90°. Upon the change in the input intensity from 1 kW/cm^2 to 4 GW/cm^2, the phase changed by 180° due to the changes in the relative spectral position of the electric and magnetic resonances.

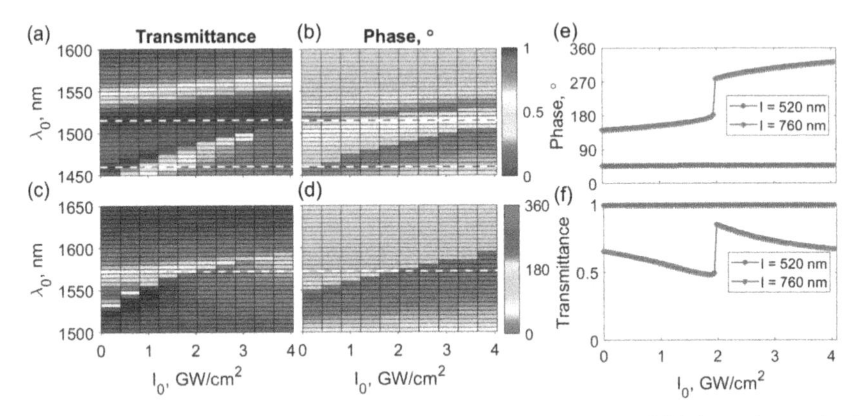

Figure 5. Nonlinear study results of ChG nano-blocks. (**a**) Transmittance and (**b**) phase of transmitted light with side-length 700 nm. (**c**) Transmittance and (**d**) phase of transmitted light with side-length 760 nm. (**e**) Transmittance and (**f**) phase of transmitted light with side-length $l = 520$ nm and $l = 760$ nm at the operation wavelength of 1572 nm.

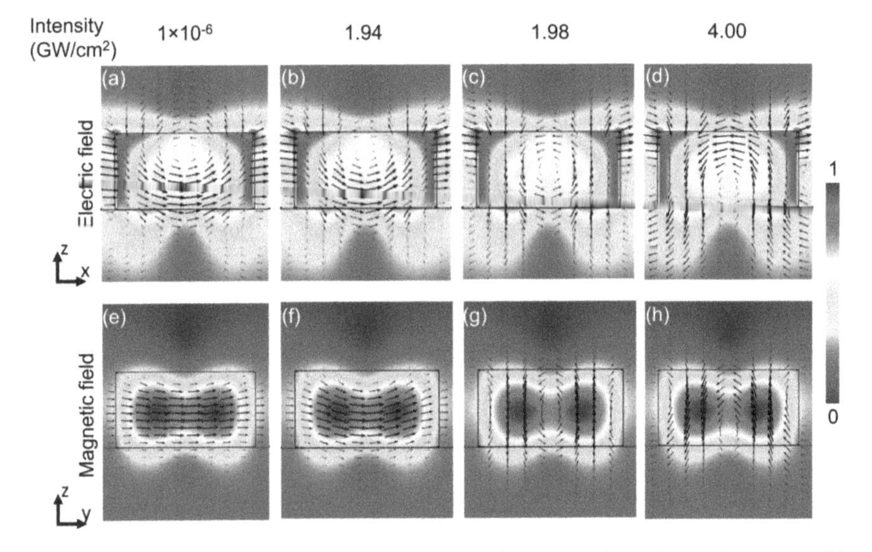

Figure 6. Normalized electric (**a–d**) and magnetic fields (**e–h**) at four selected input intensities. The input intensities are labeled on the top. Both the arrow length and the color maps represent the field intensity. The color plots are normalized to the maximum value on each plot. Panels (**a**,**b**,**e**,**f**) show that before the phase jump, only a magnetic resonance is present at the operation wavelength. When the intensity increases, both resonances shift to longer wavelengths; panels (**c**,**d**,**g**,**h**) show that for intensity larger than 1.98 GW/cm^2 both electric and magnetic resonances are present at the operation wavelength, which results in the 90° phase jump.

To verify the ability to produce the OAM beam with positive or negative topological charge, a metasurface with four quadrants and 50×50 unit cells in each quadrant was simulated using the CST Microwave Studio time domain solver and the propagating of the resulting near-field distribution in free space for 4 mm is simulated using the beam propagation method [38]. Figure 7a shows the schematic of the metasurface with an HG input beam normally incident from the substrate side. The intensity and phase distribution of the transmitted beam in low intensity case are shown in

Figure 7b,c, respectively. The dark center in Figure 7b and the spiral phase distribution in Figure 7c show that the OAM is carried by the transmitted beam. In this case, the refractive index of ChG is $n_{ChG} = n_0 = 2.43$. To simulate the nonlinear metasurface, the refractive index of ChG is set as $n_{ChG} = n_0 + \Delta n = 2.53$. Therefore, the phase acquired in quadrants II and IV of the metasurface changed by 180°. The light transmitted in the nonlinear regime, shown in Figure 7d,e, also has a singularity in the center, but the direction of the spiral phase distribution is opposite to Figure 7c, which means it carries an OAM with the opposite sign.

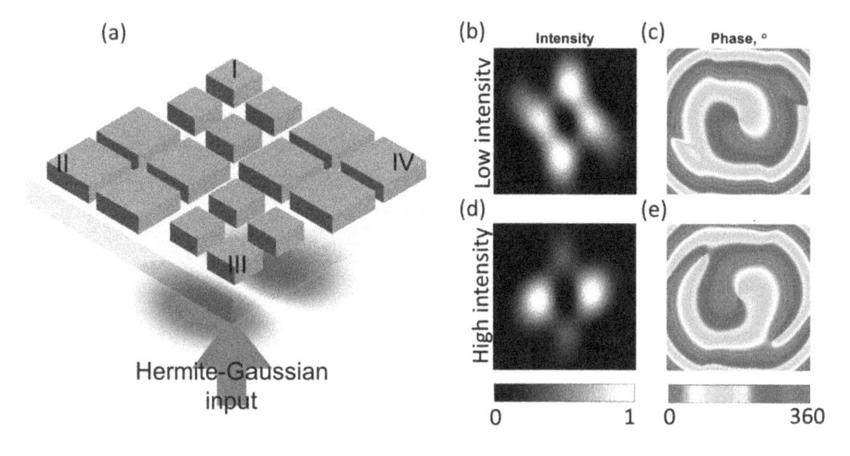

Figure 7. Schematic and simulation results of the nonlinear tunable metasurface. (**a**) Schematic of a metasurface consisting of four quadrants with an HG input beam incident from the bottom. (**b**) Intensity and (**c**) phase distribution of the light transmitted through the metasurface at low intensity. The refractive index of ChG is $n_{ChG} = n_0 = 2.43$. (**d**) Intensity and (**e**) phase distribution of the light transmitted through the metasurface at high intensity. The refractive index of ChG is assumed to be uniform and to have a value $n_{ChG} = n_0 + \Delta n = 2.53$.

3. Summary

In this work, we have experimentally demonstrated an OAM-beam-converting metasurface enabled by the Mie-resonances of the ChG cubes operating at the telecommunication wavelength. We designed and fabricated the beam converter switching an HG mode to an OAM beam. The fabrication of the ChG film requires only single-step lithography. Moreover, ChG possesses relatively large third-order nonlinearity at near-infrared wavelengths. A nonlinear metasurface which can generate reconfigurable OAM beams with opposite topological charges is designed and demonstrated theoretically. With an HG input beam, the output of the metasurface in the linear regime has a helical wavefront and carries an OAM with topological charge plus one. When the input intensity increases to a specific value in the nonlinear regime, the output wavefront is still helical but twisted in the opposite direction than in the linear regime, which leads to a negative charge of the OAM carried by the beam. The experimental realization of the nonlinear tunable metasurface might be enabled with the use of silver-doped As_2S_3, which is characterized by a nonlinear coefficient two orders of magnitude larger than that of a pure As_2S_3 film. This direction will be explored in future studies and the results will be presented elsewhere.

4. Materials and Methods

Design: we used the CST Microwave Studio Frequency Solver to design the linear metasurface. The refractive indices of materials were measured using a spectroscopic ellipsometer. The nonlinear simulations were performed in Comsol Multiphysics. The nonlinear coefficient was measured using a in-home Z-scan setup.

Sample fabrication: an array of square chromium windows with size 150 μm × 150 μm was fabricated on a glass substrate. Then, the As_2S_3 film with the thickness of 500 nm was deposited on top via thermal evaporation in a Lesker PVD 75 deposition system equipped with a low temperature evaporation source. During the deposition, the substrate temperature was maintained at approximately 20 °C. Inside each of the Cr windows, the ChG was patterned with a square array of blocks using electron-beam lithography (Vistec EBPG5000+ 100KV) with dosage 10.5 mC/cm^2. The area of the pattern is 132 μm × 132 μm. The chromium around the sample played a double role: (i) it enhanced the reflectivity of the substrate enabling automatic sample alignment, and (ii) it increased the sample conductivity allowing us to avoid accumulation of charges. After exposure, the sample was immersed in a 1:1 mixture of Microposit MF-319 developer and deionized water for 32 seconds to develop. After the development, the parts unexposed by the electron beam have been removed and the thickness of the sample was found to be 400 nm using the atomic-force microscopy. Finally, six layers of Poly(methyl methacrylate) were spin-coated on the sample to provide a symmetric refractive index.

Experiment: to characterize the fabricated metasurface, we built a Mach–Zehnder interferometer, as shown in Figure 4e. To generate an HG mode, a phase plate was inserted in the main beam path. A glass substrate spin-coated with a layer of S1813 photoresist was used as a phase plate to delay the one part of the beam by half of the wavelength. The photoresist on half of the substrate was removed using photolithography. The edge of the photoresist was placed at the center of the main beam and a spatial light filter was placed after the phase plate to filter the HG mode.

Data availability: the data that support the findings of this study are available from the authors on reasonable request, see author contributions for specific data sets.

Author Contributions: Y.X., J.S. and N.M.L. contributed to the initial idea. Y.X., J.S., M.I.S., W.W. and A.P. performed the numerical simulation. [illegible] prepared the As_2S_3 ChG film. Y.X., J.S., M.I.S. and A.T. patterned the nanostructures on the samples. Y.X., J.S., M.I.S., W.W. and N.M.L. wrote the manuscript. N.M.L. supervised the work.

Funding: This research was funded by U.S. Office of Naval Research (N000141613020).

Conflicts of Interest: The authors declare no conflict of interest.

Abbreviations

The following abbreviations are used in this manuscript:

HG	Hermite–Gaussian
SLM	Spatial light modulator
SPP	Spiral phase plate
OAM	Orbital angular momentum
ChG	Chalcogenide glass

References

1. Allen, L.; Beijersbergen, M.W.; Spreeuw, R.J.C.; Woerdman, J.P. Orbital angular momentum of light and the transformation of Laguerre–Gaussian laser modes. *Phys. Rev. A* **1992**, *45*, 8185–8189. [CrossRef] [PubMed]
2. Tamburini, F.; Mari, E.; Sponselli, A.; Thidé, B.; Bianchini, A.; Romanato, F. Encoding many channels on the same frequency through radio vorticity: First experimental test. *New J. Phys.* **2012**, *14*, 033001. [CrossRef]
3. Wang, J.; Yang, J.Y.; Fazal, I.M.; Ahmed, N.; Yan, Y.; Huang, H.; Ren, Y.; Yue, Y.; Dolinar, S.; Tur, M.; et al. Terabit free-space data transmission employing orbital angular momentum multiplexing. *Nat. Photonics* **2012**, *6*, 488–496. [CrossRef]
4. Willner, A.E.; Huang, H.; Yan, Y.; Ren, Y.; Ahmed, N.; Xie, G.; Bao, C.; Li, L.; Cao, Y.; Zhao, Z.; et al. Optical communications using orbital angular momentum beams. *Adv. Opt. Photonics* **2015**, *7*, 66–106. [CrossRef]

5. Bozinovic, N.; Yue, Y.; Ren, Y.; Tur, M.; Kristensen, P.; Huang, H.; Willner, A.E.; Ramachandran, S. Terabit-Scale Orbital Angular Momentum Mode Division Multiplexing in Fibers. *Science* **2013**, *340*, 1545–1548. [CrossRef] [PubMed]
6. Gaffoglio, R.; Cagliero, A.; Veccjo, G. Vortex Waves and Channel Capacity: Hopes and Reality. *IEEE Access* **2018**, *6*, 19814–19822. [CrossRef]
7. Padgett, M.J. Orbital angular momentum 25 years on [Invited]. *Opt. Express* **2017**, *25*, 11265–11274. [CrossRef] [PubMed]
8. Zürch, M.; Kern, C.; Hansinger, P.; Dreischuh, A.; Spielmann, C. Strong-field physics with singular light beams. *Nat. Phys.* **2012**, *10*, 743. [CrossRef]
9. Sun, J.; Silahli, S.Z.; Walasik, W.; Li, Q.; Johnson, E.; Litchinitser, N.M. Nanoscale orbital angular momentum beam instabilities in engineered nonlinear colloidal media. *Opt. Express* **2018**, *5*, 5118–5125. [CrossRef]
10. Yu, N.; Capasso, F. Flat optics with designer metasurfaces. *Nat. Mater.* **2014**, *13*, 139–150. [CrossRef]
11. Yu, N.; Genevet, P.; Kats, M.A.; Aieta, F.; Tetienne, J.P.; Capasso, F.; Gaburro, Z. Light propagation with phase discontinuities: Generalized laws of reflection and refraction. *Science* **2011**, *334*, 333–337. [CrossRef] [PubMed]
12. Kim, S.W.; Yee, K.J.; Abashin, M.; Pang, L.; Fainman, Y. Composite dielectric metasurfaces for phase control of vector field. *Opt. Lett.* **2015**, *40*, 2453–2456. [CrossRef] [PubMed]
13. Minovich, A.E.; Miroshnichenko, A.E.; Bykov, A.Y.; Murzina, T.V.; Neshev, D.N.; Kivshar, Y.S. Functional and nonlinear optical metasurfaces. *Laser Photon. Rev.* **2015**, *9*, 195–213. [CrossRef]
14. Wu, C.; Arju, N.; Kelp, G.; Fan, J.A.; Dominguez, J.; Gonzales, E.; Tutuc, E.; Brener, I.; Shvets, G. Spectrally selective chiral silicon metasurfaces based on infrared Fano resonances. *Nat. Commun.* **2014**, *5*, 3892. [CrossRef] [PubMed]
15. Ni, X.J.; Kildishev, A.V.; Shalaev, V.M. Metasurface holograms for visible light. *Nat. Commun.* **2013**, *4*, 1–6. [CrossRef]
16. Buchnev, O.; Podoliak, N.; Kaczmarek, M.; Zheludev, N.I.; Fedotov, V.A. Electrically controlled nanostructured metasurface loaded with liquid crystal: Toward multifunctional photonic switch. *Adv. Opt. Mater.* **2015**, *3*, 674–679. [CrossRef]
17. Sautter, J.; Staude, I.; Decker, M.; Rusak, E.; Neshev, D.N.; Brener, I.; Kivshar, Y.S. Active tuning of all-dielectric metasurfaces. *ACS Nano* **2015**, *9*, 4308–4315. [CrossRef] [PubMed]
18. Pfeiffer, C.; Emani, N.K.; Shaltout, A.M.; Boltasseva, A.; Shalaev, V.M.; Grbic, A. Efficient light bending with isotropic metamaterial Huygens' surfaces. *Nano Lett.* **2014**, *14*, 2491–2497. [CrossRef]
19. Holloway, C.L.; Kuester, E.F.; Gordon, J.A.; O'Hara, J.; Booth, J.; Smith, D.R. An overview of the theory and applications of metasurfaces: The two-dimensional equivalents of metamaterials. *IEEE Antennas Propag. Mag.* **2012**, *54*, 10–35. [CrossRef]
20. Lin, D.; Fan, P.; Hasman, E.; Brongersma, M.L. Dielectric gradient metasurface optical elements. *Science* **2014**, *345*, 298–302. [CrossRef]
21. Yin, X.B.; Ye, Z.L.; Rho, J.; Wang, Y.; Zhang, X. Photonic spin Hall effect at metasurfaces. *Science* **2013**, *339*, 1405–1407. [CrossRef] [PubMed]
22. Karimi, E.; Schulz, S.A.; De Leon, I.; Qassim, H.; Upham, J.; Boyd, R.W. Generating optical orbital angular momentum at visible wavelengths using a plasmonic metasurface. *Light Sci. Appl.* **2014**, *2*, e167. [CrossRef]
23. Zhao, Q.; Zhou, J.; Zhang, F.; Lippens, D. Mie resonance-based dielectric metamaterials. *Mater. Today* **2009**, *12*, 60–69. [CrossRef]
24. Yao, Y.; Shankar, R.; Kats, M.A.; Song, Y.; Kong, J.; Loncar, M.; Capasso, F. Electrically tunable metasurface perfect absorbers for ultrathin mid-infrared optical modulators. *Nano Lett.* **2014**, *14*, 6526–6532. [CrossRef] [PubMed]
25. Shalaev, M.I.; Sun, J.; Tsukernik, A.; Pandey, A.; Nikolskiy, K.; Litchinitser, N.M. High-Efficiency All-Dielectric Metasurfaces for Ultracompact Beam Manipulation in Transmission Mode. *Nano Lett.* **2015**, *15*, 6261–6266. [CrossRef] [PubMed]
26. Fu, Y.H.; Kuznetsov, A.I.; Miroshnichenko, A.E.; Yu, Y.F.; Luk'yanchuk, B. Directional visible light scattering by silicon nanoparticles. *Nat. Commun.* **2013**, *4*, 1527. [CrossRef] [PubMed]
27. Lapine, M.; Shadrivov, I.V.; Kivshar, Y.S. Colloquium: Nonlinear metamaterials. *Rev. Mod. Phys.* **2014**, *86*, 1093. [CrossRef]

28. Shadrivov, I.V.; Kapitanova, P.V.; Maslovski, S.I.; Kivshar, Y.S. Metamaterials controlled with light. *Phys. Rev. Lett.* **2012**, *109*, 083902. [CrossRef]
29. Pandey, A.; Litchinitser, N.M. Nonlinear light concentrators. *Opt. Lett.* **2012**, *37*, 5238–5240. [CrossRef]
30. Shcherbakov, M.R.; Vabishchevich, P.P.; Shorokhov, A.S.; Chong, K.E.; Choi, D.Y.; Staude, I.; Miroshnichenko, A.E.; Neshev, D.N.; Fedyanin, A.A.; Kivshar, Y.S. Ultrafast all-optical switching with magnetic resonances in nonlinear dielectric nanostructures. *Nano Lett.* **2015**, *15*, 6985–6990. [CrossRef]
31. Shcherbakov, M.R.; Liu, S.; Zubyuk, V.V.; Vaskin, A.; Vabishchevich, P.P.; Keeler, G.; Pertsch, T.; Dolgova, T.V.; Staude, I.; Brener, I.; et al. Ultrafast all-optical tuning of direct-gap semiconductor metasurfaces. *Nat. Commun.* **2017**, *8*, 17. [CrossRef] [PubMed]
32. Xu, Y.; Sun, J.; Frantz, J.; Shalaev, M.I.; Walasik, W.; Pandey, A.; Myers, J.D.; Bekele, R.Y.; Tsukernik, A.; Sanghera, J.S.; et al. Reconfiguring structured light beams using nonlinear metasurfaces. *Opt. Express* **2018**, *23*, 30930–30943. [CrossRef] [PubMed]
33. Sanghera, J.S.; Shaw, L.B.; Pureza, P.; Nguyen, V.Q.; Gibson, D.; Busse, L.; Aggarwal, I.D.; Florea, C.M.; Kung, F.H. Nonlinear properties of chalcogenide glass fibers. *Int. J. Appl. Glass SCI* **2010**, *1*, 296–308. [CrossRef]
34. Eggleton, B.J.; Luther–Davies, B.; Richardson, K. Chalcogenide photonics. *Nat. Photonics* **2011**, *5*, 141–148. [CrossRef]
35. Hilton, A.R.; Kemp, S. *Chalcogenide Glasses for Infra-Red Optics*; McGraw Hill: New York, NY, USA, 2010.
36. Vlcek, M.; Jain, H. Nanostructuring of chalcogenide glasses using electron beam lithography. *J. Optoelectron. Adv. Mater.* **2006**, *8*, 2108–2111.
37. Van de Groep, J.; Polman, A. Designing dielectric resonators on substrates: Combining magnetic and electric resonances. *Opt. Express* **2013**, *21*, 26285–26302. [CrossRef]
38. Feit, M.D.; Fleck, J.A. Light propagation in graded-index optical fibers. *Appl. Opt.* **1978**, *17*, 3990–3998. [CrossRef]
39. Zhang, M.; Li, T.; Yang, Y.; Tao, H.; Zhang, X.; Yuan, X.; Yang, Z. Femtosecond laser induced damage on Ge–As–S chalcogenide glasses. *Opt. Mater. Express* **2019**, *2*, 352213. [CrossRef]
40. You, C.; Dai, S.; Zhang, P.; Xu, Y.; Wang, Y.; Xu, D.; Wang, R. Mid-infrared femtosecond laserinduced damages in As_2S_3 and As_2Se_3 chalcogenide glasses. *Sci. Rep.* **2017**, *7*, 6497. [CrossRef]
41. Kosa, T.I.; Rangel–Rojo, R.; Hajto, E.; Ewen, P.J.S.; Owen, A.E.; Kar, A.K.; Wherrett, B.S. Nonlinear optical properties of silver-doped As_2S_3. *J. Non-Cryst. Solids* **1993**, *164*, 1219–1222. [CrossRef]

Communication

Reception of OAM Radio Waves Using Pseudo-Doppler Interpolation Techniques: A Frequency-Domain Approach

Marek Klemes

Canada Research Center, Huawei Technologies Canada Co. Ltd., 303 Terry Fox Drive, Kanata, ON K2K 3J1, Canada; marek.klemes@huawei.com

Received: 31 January 2019; Accepted: 10 March 2019; Published: 14 March 2019

Abstract: This paper presents a practical method of receiving waves having orbital angular momentum (OAM) in the far field of an antenna transmitting multiple OAM modes, each carrying a separate data stream at the same radio frequency (RF). The OAM modes are made to overlap by design of the transmitting antenna structure. They are simultaneously received at a known far-field distance using a minimum of two antennas separated by a short distance tangential to the OAM conical beams' maxima and endowed with different pseudo-Doppler frequency shifts by a modulating arrangement that dynamically interpolates their phases between the two receiving antennas. Subsequently down-converted harmonics of the pseudo-Doppler shifted spectra are linearly combined by sets of weighting coefficients which effectively separate each OAM mode in the frequency domain, resulting in a higher signal-to-noise ratios (SNR) than possible using spatial-domain OAM reception techniques. Moreover, no more than two receiving antennas are necessary to separate any number of OAM modes in principle, unlike conventional MIMO (Multi-Input, Multi-Output) which requires at least K antennas to resolve K spatial modes.

Keywords: orbital angular momentum; phase mode; twisted waves; radio frequency; receiver; pseudo-Doppler; interpolation; multi-input multi-output; MIMO; frequency-domain; time-gated frequency-shift interpolation

---

1. Introduction

Since 1992, much effort has been devoted to the exploitation of the property of waves called orbital angular momentum (OAM). Although it appears to be newly-appropriated from the physics community, OAM has been known previously, especially in the RF (Radio Frequency) community, as phase modes. Phase modes were useful for synthesizing excitations of circular arrays in radio direction-finding and null-steering applications since the 1960's [1–4]. Since the advent of multi-input multi-output (MIMO) technology in radio communications, OAM came to be recognized as another spatial dimension to be exploited for enhancing capacity of radio communications, and also optical communications in free space as well as fiber. An excellent historical summary of the development of OAM applications is given in [5].

No shortage of literature exists about how to generate and characterize OAM radio and optical waves [6]. Relatively few investigations focus on applications in radio communications, which is our interest in this paper, along the lines of [7,8].

Even fewer investigations focus on the receiving end of OAM communications links, with most of them relying on spatial techniques employing the same principles as those for generating the OAM modes at the transmitting end.

Consequently, most attempts at exploiting the OAM modes to enhance capacity of radio links suffer from the limitations imposed upon the receiver and antennas due to the spatial minima of all nonzero-order OAM modes' beams on their axes, in the far field beyond the Rayleigh distance [7], where the receiver is situated. These limitations lead to low signal-to-noise ratios (SNRs), or very large receiving antennas, limitation to very short wavelengths, or limitations to short ranges comparable to the Rayleigh distance, or some combination of these. Additionally, real-world effects of antenna imperfections, multipath and dispersive propagation only add to the difficulties of reliably receiving and resolving the OAM modes and extracting the information streams from them.

Of all the more than 300 references cited in [6], only two [9,10] are about methods of reception and resolution of OAM modes that do not rely on variations of the spatial matched-filter concept whereby the helical phase fronts of the OAM beams are "untwisted" back to planar ones by the receiving antenna structure.

In this paper, a variation of the method of [10] is pursued to show mathematically, and via simulation, how it can be applied to the transmission and reception of multiple OAM modes simultaneously without incurring most of the limitations of prior methods cited above. A patent-pending apparatus for realizing this method at a radio receiver in real-time is also described, coupled with a signal-processing algorithm to resolve the different data streams carried on the OAM beams.

Section 2 presents the background and pseudo-Doppler principles behind the "virtual rotational antenna" of [10] and relates it to the context of radio beams possessing OAM. Transmission of OAM-bearing radio beams is also briefly reviewed. Section 3 presents the expression for the received OAM radio signals at the output of the pseudo-Doppler modulated antenna apparatus and relates its parameters to the physical geometry of the antennas and radio link. Section 4 describes the signal processing algorithm which processes the pseudo-Doppler aggregate signal to resolve the data streams carried on the individual OAM modes and examines some of its variants and limitations. It also contains preliminary simulation results to support the analyses. Conclusions and directions for future work are presented in Section 5. Select mathematical details are contained in the appendices.

2. Background and Pseudo-Doppler Principles

2.1. Real and Virtual Doppler Effect

Because most of the analyses involve circular geometry, it is useful to proceed in those terms. Accordingly, visualize an antenna element at position "n" on a circular locus having radius R as in Figure 1, with a plane wave incident on it from a point source "P" at distance L from the center of the circle, in the far field. At the antenna element the phase of the incident plane wave relative to that at the source is given by

$$\varphi_n = \frac{2\pi}{\lambda}(L - R\cos(\theta - \theta_n)\sin\phi) \tag{1}$$

Next, imagine that the antenna is moving along the circular locus with tangential velocity, v, in the direction of the colored arrow. With the radius remaining constant at R, this velocity involves only the change in azimuth angle θ with time, as $v = Rd\theta/dt$. The corresponding change in phase at the antenna is derived by applying the chain rule as

$$\frac{d\varphi_n}{dt} = \frac{2\pi v}{\lambda}(\sin(\theta - \theta_n)\sin\phi) \tag{2}$$

where $f_{Doppler} = v/\lambda$ is recognized as the Doppler shift frequency due to the tangential motion at velocity v. This frequency is imposed upon the signal received at the antenna and if its angular position varies uniformly in time as $\theta(t)$, it results in a sinusoidal frequency modulation of the received signal with a deviation equal to $f_{Doppler}$ (when elevation angle is $\varphi = \pi/2$) and phase corresponding to the azimuth direction of arrival of the plane wave. This is a first hint that spatial information about a received signal can be determined in frequency domain. This phenomenon is sometimes used in radio azimuth direction-finding applications.

The motion of the antenna does not have to be real—it can be emulated using several antennas spaced at intervals of d around the circular locus and switching their outputs to the analysis receiver at intervals of $\tau = d/v$. It then appears to the analysis receiver that it is sampling the output of one antenna moving around the locus with velocity v at intervals of τ, because it observes the same Doppler shift, which is actually a pseudo-Doppler shift as there is no real motion involved.

In the present application any actual motion of the source or the receiving antennas will be assumed to be 0, and the source of the OAM signals will be positioned at elevation angle $\varphi = 0$ to keep the analysis simple. The reference phase can be taken at the center of the circle in the far field by setting $L = 0$. The only relevant phase shifts will then be relative phases between two or more antenna elements on the same locus.

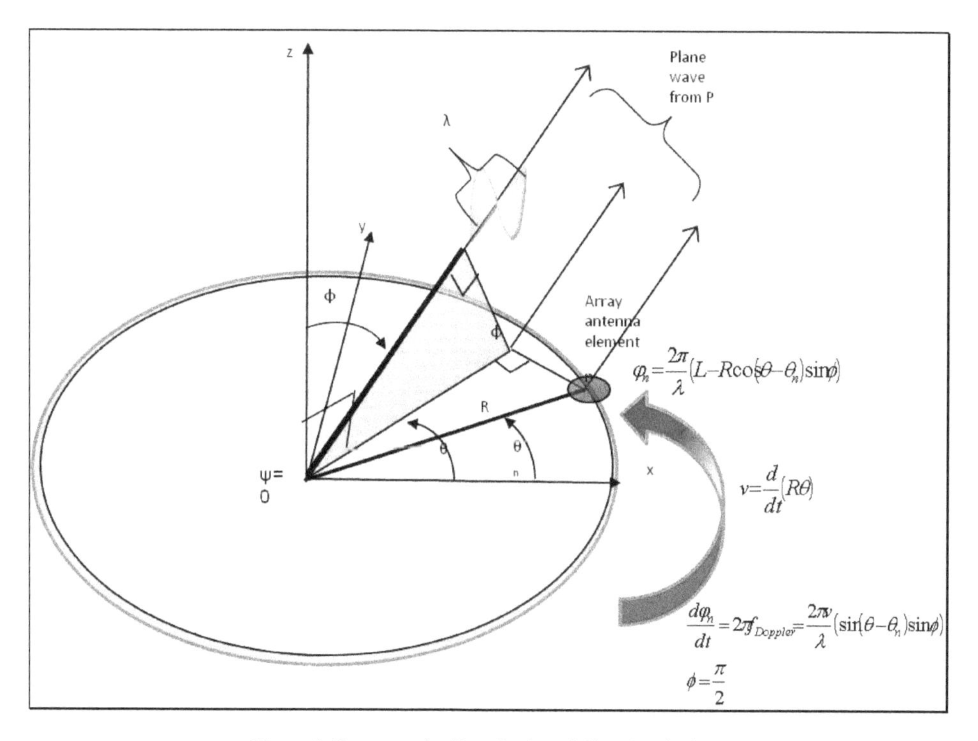

Figure 1. Geometry for Doppler-based direction-finding.

2.2. Application to OAM Radio Waves in the Far Field

It is instructive to review the salient features of radio waves possessing various orders of OAM, which will be denoted by integers $\pm k$. Such radio waves are generated by imposing a phase shift of $k2\pi$ radians for every revolution of the observation point around the beam axis, giving it a helical phase front. This is not to be confused with polarization, which can be of any type. In RF applications, this can be relatively easily achieved using a uniform circular array of K identical antenna elements, each one fed by a current that is shifted in phase from that of its neighbor (in one direction) by $k2\pi/K$ radians and with the same amplitude. Negative phase shifts generate OAM modes with helical phase fronts winding in the opposite sense around the beam axis, up to order $K/2$-1.

A common method of creating multiple OAM beam excitations of the same circular array of antenna elements is to connect the K elements to the K output ports of a modified Butler matrix, and the K input ports of that Butler Matrix to K transmitters in the same RF band, with each modulated by a different stream of independent data symbols. The Butler matrix must be modified so as to possess one port which gives rise to a zero-th order OAM mode; otherwise the electrical phases at the elements do not progress through an integer number of cycles so the phase fronts would not form continuous spirals.

When plotted in three dimensions, the beam patterns appear conical for all non-zero orders of OAM, as depicted in Figure 2, where color was used to denote the electrical phase at a fixed time, modulo-2π radians.

It is instructive to note that the phase (color) patterns rotate around the beam axis at the RF rate in time, i.e., one revolution per cycle of the radio frequency. Therefore, k-phase fronts (of a given color) pass a point on the cone of the k-th OAM beam in the tangential direction, per period of the RF carrier wave. Equivalently, at any given point in time, an electrical phase gradient of $k2\pi/(2\pi R)$ radians per meter exists along the circular locus (also the beam footprint) around the axis of the conical beam of the k-th OAM mode.

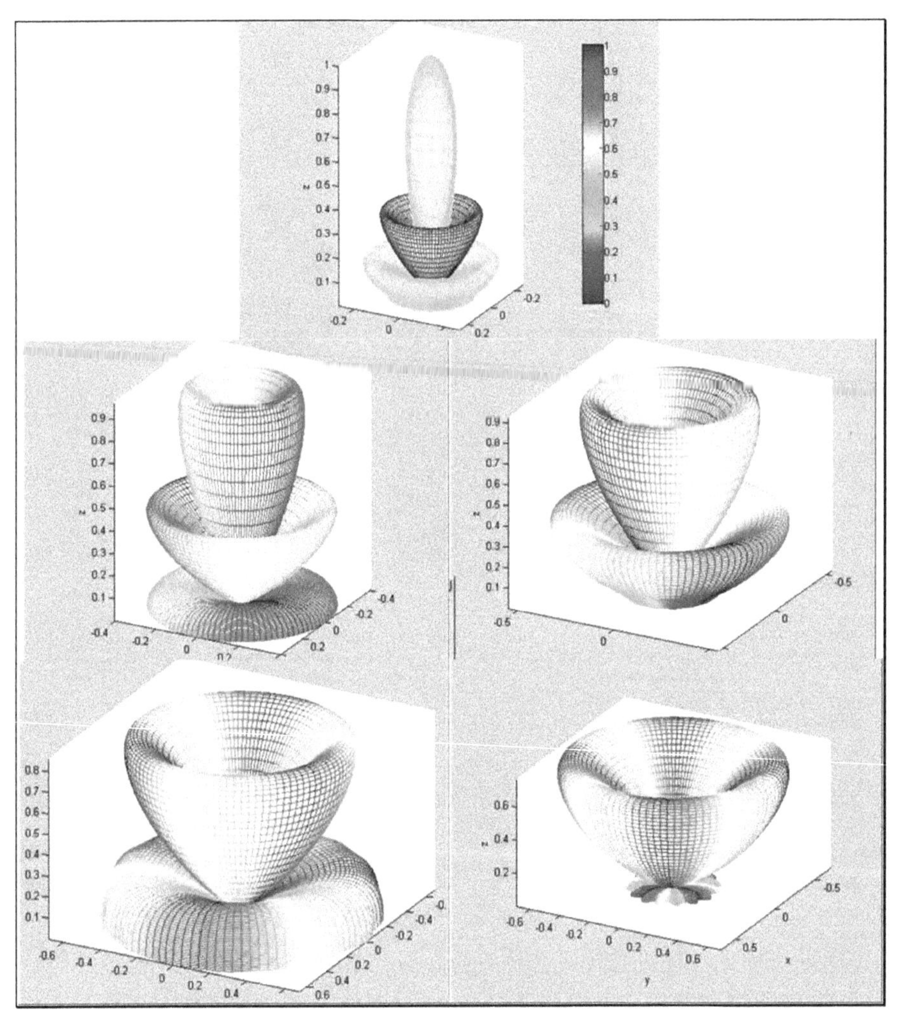

Figure 2. Far-field beam patterns of OAM modes $k = 0$ (top), 1, 2, 3, 4.

In Figure 2, the antenna elements in the x-y plane numbered $K = 16$ and were omnidirectional, with the beam axes being in the z (vertical) direction.

With the help of Figure 3, visualize the circular locus of the antenna element in Figure 1 as coinciding with the peak of a conical beam of OAM order k, whose source is a circular array in the far field on the z-axis.

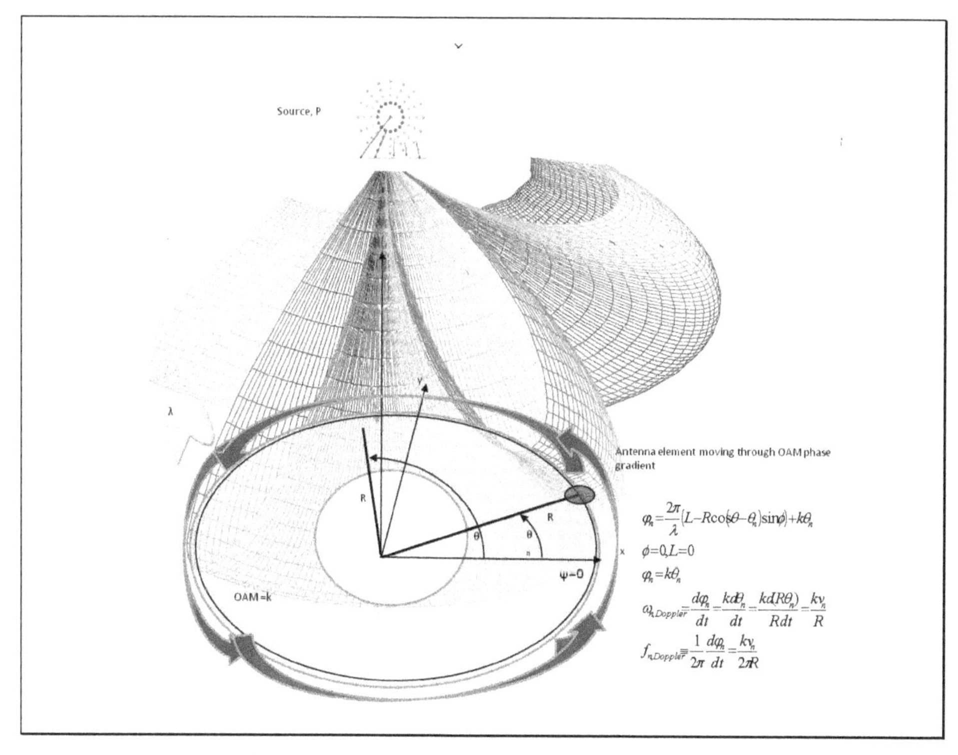

Figure 3. Receiving antenna moving through OAM beam.

Because the elevation angle of the source is now $\varphi = 0$, the only phase variation along the locus of the antenna element at a given point in time is due to that of the OAM beam. The shaded circle denotes the footprint area of the k-th OAM beam.

Taking the phase at the x-axis as the reference phase of the moving antenna element, its phase at position θ_n is therefore given simply by

$$\varphi_n = k\theta_n \tag{3}$$

at a given point in time. As the antenna moves around the circular locus in the x-y plane with uniform velocity v, its angular position changes linearly with time, consequently causing its electrical phase to vary linearly with time according to (3) as

$$\frac{d\varphi_n}{dt} = \frac{kd\theta_n}{dt} \tag{4}$$

This can be related to a kind of "transverse" Doppler frequency shift because with R being constant, (4) can be written also as

$$\frac{d\varphi_n}{dt} = \frac{kd\theta_n}{dt} = \frac{kv_n}{R} = 2\pi f_{n,Doppler} \tag{5}$$

since, according to the discussion of Figure 1, it is clear that

$$v_n = \frac{d}{dt}(R\theta_n) \tag{6}$$

Thus, it has been shown that a spatial-domain property of an OAM beam of order k, the phase gradient k/R, can be converted to a frequency-domain property, namely a kind of transverse or rotational Doppler shift $f_{n,Doppler}$, through the motion of the antenna element receiving the OAM beam. Note that the effect is real in the physical sense [9]; the subscript "n" may be omitted as there is only the one moving antenna. Note also, that this transverse Doppler shift is directly proportional to the OAM order, k, and independent of RF carrier frequency.

Next, invoke the pseudo-Doppler technique whereby the motion of a single antenna from position #1 to position #2 is emulated by switching among several antennas, as outlined at the end of subsection *A*. Specifically, let the receiver employ two antennas separated by distance "d" tangentially to the footprint of the OAM beam, and instead of switching between their outputs, the receiver *combines* their outputs in *time-varying proportions* ranging from only output #1, to half of each output #1 and #2, to only output #2. This is in effect a form of gradual switching between the antennas, in *one* direction; it is assumed to be repeated periodically at some rate in accordance with the principles employed in [10].

To understand how such time-variant combining emulates a transverse Doppler shift in the received carrier frequency, Appendix A reviews the principle of the Doppler effect in simple contexts. With that in mind, it is relatively easy to derive the **key relation** between the phases of the OAM beam as received at the two antennas at a given time in Figure 4.

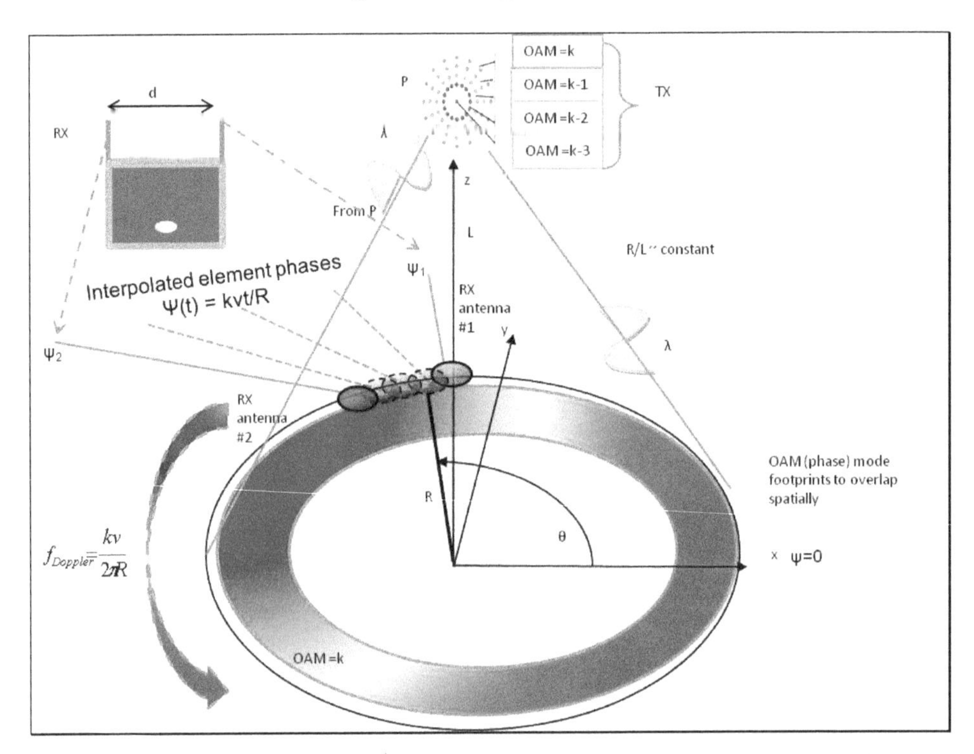

Figure 4. Applying pseudo-Doppler technique to OAM beams at a two-antenna receiver. Only the footprint of the OAM beam is shown; the color denotes the phase.

At a given point in time, the phase of the RF wave arriving in the form of OAM mode k at RX antenna #1 is ψ_1 and at the same time the phase at RX antenna #2 is ψ_2. Because the phase delay advances k multiples of 2π radians for one complete trip around the footprint, $2\pi R$, at a given observation point in time, it advances by proportion $kd/(2\pi R)$ for the portion of the footprint covered by the antenna separation "d". The phases at the two RX antennas are therefore related as

$$\psi_2(t) = \psi_1(t) - k2\pi\left(\frac{d}{2\pi R}\right) \tag{7}$$

Consequently, the signals at the inputs W_1 and W_2 of the time-varying combiner are modeled as being multiplied by the complex-exponential phase factors as

$$\begin{aligned} W_{1,k}(t) &= S_k(t)e^{j\psi_1(t)} \\ W_{2,k}(t) &= S_k(t)e^{j\psi_2(t)} = S_k(t)e^{j\psi_1(t)-jkd/R} \end{aligned} \tag{8a,b}$$

where $S_k(t)$ is the signal of the k-th OAM beam received at the reference point in the far field.

Before exposing the function of the time-varying combiner in Section 3, the next subsection describes briefly the transmitting end of the link where the multiple OAM beams are modulated with independent data streams on the same RF carrier and launched from the antenna structure. In this respect, the multiplexing of several data streams onto several OAM beams is still effectively performed in the spatial domain as in all other OAM transmission schemes in RF applications.

2.3. *Transmission of Multiple Overlapping OAM Radio Beams*

In numerous past applications, the axial beam patterns of phase modes, or OAM modes in modern parlance, have been derived and characterized as being proportional to [4]

$$G_k\left(\theta, \phi\right) \approx (-j)^k e^{jk\theta} J_k((2\pi r \sin\phi)/\lambda) \tag{9}$$

where J_k is a Bessel function of the first kind, order k, r is the radius of the circular antenna array, φ is the elevation angle measure from the beam axis and λ is the wavelength of the RF carrier wave of the k-th OAM mode. The first few orders of this Bessel function are shown plotted in Figure A3 of Appendix B.

Consequently, it is seen that OAM beams of higher orders have wider cone angles in the far field than those of lower orders, and beams of different OAM orders do not overlap much in space (except negative and positive modes of the same order). That is also evident from Figure 1, which was plotted in accordance with Equation (9) and Appendix B, where the peak positions along the x-axis correspond the peaks of the conical OAM beams at radii "R" from the beam axis. Clearly, the higher-order Bessel functions having peaks at larger values of "x" means that higher-order OAM modes have peaks at larger radii from the axis, hence larger cone angles, as dictated by (9). Also evident from (9) is the property that an OAM beam of order k generated from an array with a larger radius will have a smaller cone angle than the same OAM of order k generated by an array with a smaller radius.

Because it is desired to transmit multiple data streams on the same RF carrier to one user using multiple OAM beams, their conical beam patterns must overlap at the user's location in the far field. Therefore, all OAM modes cannot be launched from the same circular array of antenna elements, but the lower-order ones should be launched from arrays having proportionally smaller radii and the higher-order ones from arrays having proportionally larger radii. These arrays may be concentrically stacked as shown for example in Figure 1a of [11], adapted below as Figure 5. Such a transmitting antenna arrangement is expected to be of the same physical size as in other, more "conventional" RF schemes for transmitting OAM beam and no attempt to improve the link SNR is inferred here; that is effected at the receiver—as will be shown in subsequent sections.

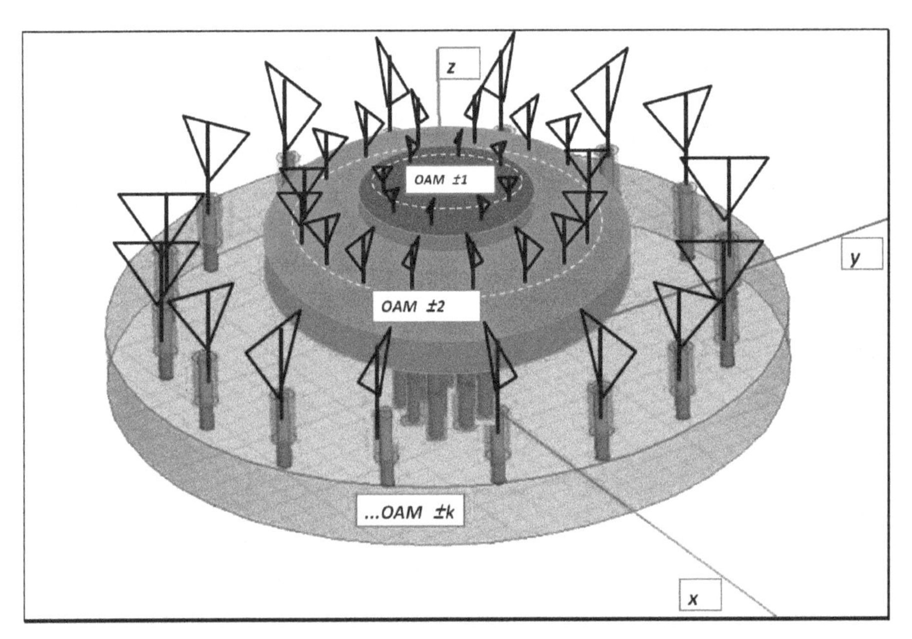

Figure 5. Example of stacked circular transmitting arrays for OAM multiplexing, after [11].

3. Real-Time Implementation of Pseudo-Doppler Effect on Received OAM Beams by Dynamic Antenna Combining

As noted in the Introduction, very few attempts at OAM multiplexing in electromagnetic-wave communication links did not rely on spatial rectification of the helical OAM phase fronts at the receiver, but a few should be mentioned before proceeding to describe the present scheme as being unique.

Reference [12] even uses a time-based method for generating OAM modes in circular arrays, which saves some hardware but otherwise is not really necessary for overcoming difficulties in implementing an OAM communications link. A simple phase-gradient measurement is used in [13] to sequentially detect OAM modes, which are sequentially encoded with data symbols at the transmitter. The authors of [14] also use the phase gradient to resolve the OAM modes at the receiver, by switching between multiple pairs of receiving antenna elements.

In [15], the authors use the phase gradient sensed by switching between two receiving elements to identify the transmitted OAM mode, which corresponds to an encoded data symbol. This generates harmonics of the switching frequency, which are used to detect the (sequentially) transmitted data symbol. No dynamic combining effects are employed, and the data symbols are transmitted and decoded sequentially.

Another variation of the time-gated generation and detection of OAM modes appears in [16], using strategically-placed antenna elements covering only part of the circular aperture that an array such as the one in [11] would utilize. The results appear rather stochastic, with relatively high cross-talk among the detected OAM modes.

A partial-circle aperture approach was also used by the authors of [17] to avoid the size issue with "conventional" OAM receiving antenna arrays. Judicious selection of the fraction of circle covered by the receiving array renders the received OAM modes orthogonal at the receive array, thus allowing them to be resolved and independently decoded. A variation on the time-switched array method of generating OAM modes using sinusoidal modulators instead of switches at the elements of a circular array is described in [18]. It is not applicable, nor easily convertible to receiving OAM modes.

In this work the driving interest is to explore ways of implementing the method advanced in [10] in real-time, to realize a more practical OAM receiver than has been possible using co-axial spatial receiving techniques based on circular antenna arrays.

In the process, it became evident that important details of the pseudo-Doppler technique are not derived with sufficient persuasion for this author, so another aim of this work is to fill in the mathematical details that allow an actual OAM radio communications link to be conceived and simulated.

Specifically, it was noted that an actual demonstration of a real-time pseudo-Doppler-shifted spectrum of the received OAM signal carrying a useful data rate was not documented in the relevant literature; in [9] the spectrum shifts shown were caused by real Doppler shift due to physical rotation of the antenna, and in [10] the spectra were obtained by off-line post-processing of rather narrow-band data. The supplementary material shows ideal sketches of the shifted spectra and also spectral shifts due to rotation of the transmitting antenna; actual demodulated data is not represented, as that was an aim of future research stated in [10].

3.1. Using a Quadrature RF Oscillator and Mixers

Without belaboring the details, a way of implementing the relative dynamic weighting of the two receiving antenna signals comes to mind using orthogonal sinusoidal modulations, visualized as in Figure 6. It is even simpler to implement than an image-rejecting mixer in the front end of many common microwave radio receivers. Note that the sinusoidal wave generator can have a very high rate, as microwave oscillators are very common and straight-forward to implement. (This high rate will be necessary to separate the OAM signals in frequency domain, thereby facilitating their signal recovery. A real Doppler shift of such a frequency would require physical motion at speeds approaching the speed of light.)

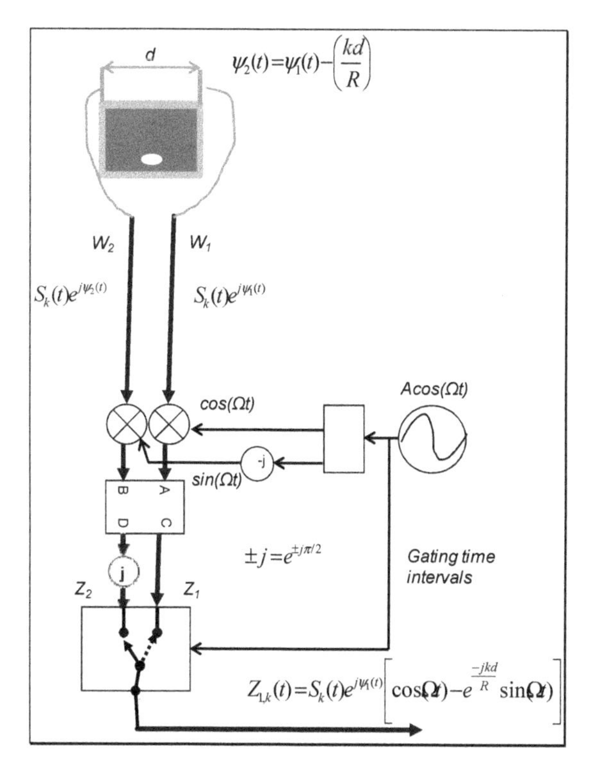

Figure 6. An alternative implementation of pseudo-Doppler OAM receiving front end.

By strategically working through the mathematics of the output equation in Figure 6, the result for the output during selected time-gated intervals can be obtained approximately as

$$\begin{aligned} Z_{1,k}(t) &\approx \sqrt{2}S_k(t)e^{j(\psi_1(t)-\frac{kd}{2R})} \times \cdots \\ \cdots &\times (\cos(\Omega t+\pi/4))e^{j(\frac{kd}{2R})(\Omega t+\pi/4)} \end{aligned} \tag{10}$$

with the understanding that $\Omega = 2\pi F$ is the radian pseudo-Doppler frequency, F being the corresponding frequency in Hz. Note that the bottom factor is due purely to the effect of the pseudo-Doppler modulator on the OAM incident wave, and the scaled pseudo-Doppler radian frequency shift $k\Omega d/(2R)$ is independent of the RF carrier radian frequency ω, which is implicit in $S(t)$, which in turn is the transmitted signal on the k-th OAM mode,

$$S_k(t) = m_k(t)e^{j\omega t} \tag{11}$$

with $m(t)$ being the modulating signal and ω being the radian carrier frequency.

Before developing the models for the necessary time-gating and demodulation functions, it is instructive to estimate the potential performance of this method of OAM reception in terms of link and antenna geometry.

3.2. Implications in an OAM Radio Link

As pointed out in [10], the resulting scaled pseudo-Doppler shift at the output of the combiner in the front end of the receiver should be greater than the bandwidth, B, of the transmitted signal in any OAM mode (assuming each OAM mode carries an independent data stream at the same rate of B symbols/second). So one "unit" of frequency shift corresponds to $k = 1$ and satisfies

$$2\pi Fd/(2R) \geq 2\pi B \tag{12}$$

in keeping with radian units of frequency. In the example depicted in Figure 7, choose the parameters as in Table 1 below. The condition $0.02k = kd/(2R) << \pi/4$ is satisfied.

Table 1. Example Parameters of OAM Radio Link.

Parameter Symbol	OAM Radio Link Parameters		
	Description	**Value**	**Units**
B	Signal bandwidth [a]	10	MHz
L	Link distance	1	km
R	Radius of overlapping OAM conical-beam footprints	20	m
d	Separation of receiving antennas	20	cm
λ	RF carrier wavelength	5	cm

[a] In each transmitted OAM mode.

From (22) and Table 1 the oscillator frequency for the pseudo-Doppler modulator of the receiver front end in Figure 7 is determined to be $F = 2$ GHz. This is a typical frequency in modern mobile radio hardware so the modulator and oscillator are easily achieved. The effective pseudo-Doppler shift of the $k= 4$ OAM mode will be 40 MHz.

Note that the receiving antenna array in Figure 7 is not positioned on the beam axis as in "conventional" RF links employing OAM beams, but at the peak of the OAM beams which is at a radius R perpendicular to the axis. It is this arrangement which enables a higher SNR at the receiving end, owing to the much higher OAM signal amplitude there.

Given the size of RF wavelengths, it would be grossly impractical to capture all of the OAM beam energy at radius R as the receive antenna array would need to be of the same order in size. As also stated in [10], it would be equally impractical to physically rotate such an antenna to impart a Doppler shift to its output, hence the motivation to use a virtually-moving, electronically interpolated antenna, to impart the much higher Doppler frequency shifts to the OAM modes using pseudo-Doppler techniques. It is not difficult to perceive that, despite having only two receiving antennas (necessarily) positioned off-axis, the SNR at the receiver can be much higher than that obtainable from a similar size of receiving radio antenna array necessarily positioned on-axis.

Therefore, the subsequent signal processing of the pseudo-Doppler shifted OAM beam signals in the frequency domain is expected to yield much "cleaner" recovered OAM signals, at longer link distances in free-space RF applications than a conventional spatial OAM recovery technique relying on the reciprocity of launching OAM modes with a similarly-sized antenna array. (Note that even in such reciprocity-based spatial OAM recovery schemes, the receiving antenna is not large enough to capture all of the beam energy contained in the toroidal "footprint" with radius R of the OAM beams at link distance L. The SNR penalty of those schemes is not due to the ***size*** of their receiving antennas so much as it is due to their necessary ***positioning on-axis***, where all (non-zero-order) OAM modes have a deep amplitude minimum and are spatially orthogonal.) Further possibilities for enhancing the effective SNR by frequency-domain signal-processing are described in Section 4.3.

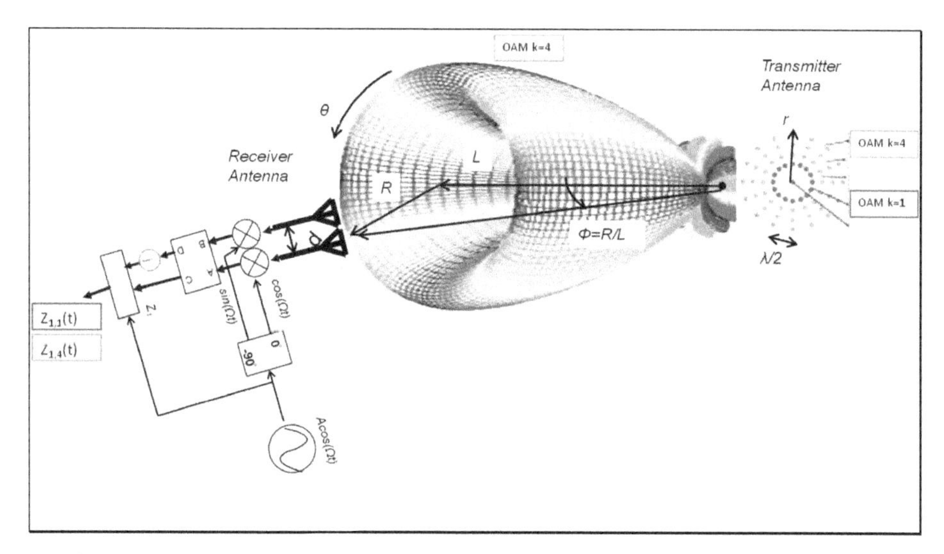

Figure 7. Example of an OAM radio link using pseudo-Doppler modulator in the off-axis receiver.

To determine the size of the transmitting antenna array, it is observed in Appendix B that the peak of the k-th order Bessel function in (9) occurs roughly where its argument is equal to $k + 1$ for orders below about 6. The receiver is situated off-axis, at the peak of the overlapping OAM beams. (The antennas are not drawn to the same scale as the beam pattern and link geometry.)

Therefore for OAM mode $k = 1$, the radius of the transmitting array is determined as

$$\frac{2\pi r_1 \sin\phi}{\lambda} = 2 \Rightarrow r_1 = \frac{2 \times 0.05\text{m}}{2\pi \times (20\text{m}/1000\text{m})} = 0.80 \text{ m}$$

and for $k = 4$, the radius of the TX circular arrays is

$$\frac{2\pi r_4 \sin\phi}{\lambda} = 5 \Rightarrow r_4 = \frac{5 \times 0.05\text{m}}{2\pi \times (20\text{m}/1000\text{m})} = 2.0 \text{ m}$$

Such size of antenna is not excessive for a sub-6 GHz base station. If the element spacing is to be half of the RF wavelength as indicated in Figure 7, then the outer array for OAM mode $k = 4$ would require 500 elements and the inner one for $k = 1$ would require 200 elements. Note that because negative OAM orders are equally handled by the same system (resulting in negative pseudo-Doppler shifts at the RX), a total of 9 OAM modes could be transmitted on this link.

4. Time Gating and Demodulation Algorithms

As noted in the discussion that relates Equation (10) to the output equation in Figure 6, the relevant approximations can be made only for certain periodic windows in time. Accordingly, the output signal at Z_1 must be gated to be observable only at those times and suppressed at all other times. It amounts to imposing time-limited frequency shift on the received signals, the frequency shifts being the pseudo-Doppler shifts of the individual superposed OAM modes. This is recognized as the frequency-domain dual of the band-limited time-delay problem, the time-delay being a fraction of the sampling interval of a discrete-time signal [19].

4.1. Time GatingTo Emulate Motion of Antenna in One Direction

The limits of the region of validity of the pseudo-Doppler frequency shift as expressed in the output by (10) are actually periodic when the fundamental period being

$$\frac{-\pi}{2} < \Omega t < 0 \tag{13}$$

in terms of phase. The output of the modulator summing junction in Figure 6, especially the bottom factor in Equation (10), must therefore be gated periodically in real time to ensure the desired frequency shifts in the final output. This periodic gating must evidently be synchronous with the pseudo-Doppler modulation, as (13) must hold for every period, or periodic values of $n\pi$.

It is instructive to visualize this synchronous gating in relation to the modulations. When superposed on the effective sinusoidal modulation waveforms applied to the antenna output signals, $\cos(\Omega t)$ to W_1 and $-\sin(\Omega t)$ to W_2, the gating intervals are seen to contain those portions of the modulations which cause one antenna output to be increasing and the other decreasing the magnitude of its contribution to the output. The alternating signs of the gating waveform based on (13) ensure that the same antenna output is always increasing while the other is decreasing. This shows that the apparatus performs the desired interpolation that emulates a moving antenna between the two stationary receive-antennas as described in [10]. Figure 8 illustrates the periodic gating of the modulation waveforms.

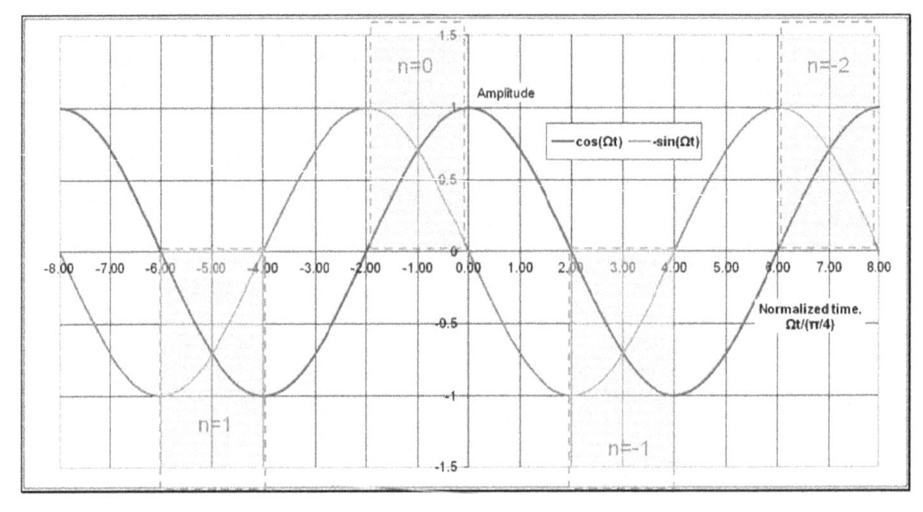

Figure 8. Antenna modulating waveforms with output gating intervals shaded.

The gating windows are denoted by the values of the period "n", and their multiplier signs by their positions either above ("+") or below ("−") the angle ("normalized time") axis. Note that the angle axis is calibrated in multiples of $\pi/4$ radians.

4.2. Simulation of Basic Pseudo-Doppler Modulator and Gating Arrangement

A preliminary numerical simulation to verify the proposed concept was conducted using MATLAB® R2012a (7.14.0.739) and Simulink® R2012a (7.9), with Communications System Toolbox (5.2), DSP System Toolbox (9.3) and Signal Processing Toolbox (6.17).

For reference, the transmitted constellation from one of the sources is observed via a matched pulse-shaping filter, as shown in Figure 9a. Its spectrum is also observed; in fact, the spectra of all the sources are the same on average and are shown in Figure 9b as seen when superposed at one receive antenna (They would appear the same on average at the other receive antenna.)

So, all the OAM modes (OAM 1 and OAM 8 in this case) occupy the same spectrum shown in Figure 9b, which is the only spectrum visible in the transmission medium and potentially subject to regulation. Yet, it will be shown that the OAM modes can be recovered separately from this composite signal in the receiver.

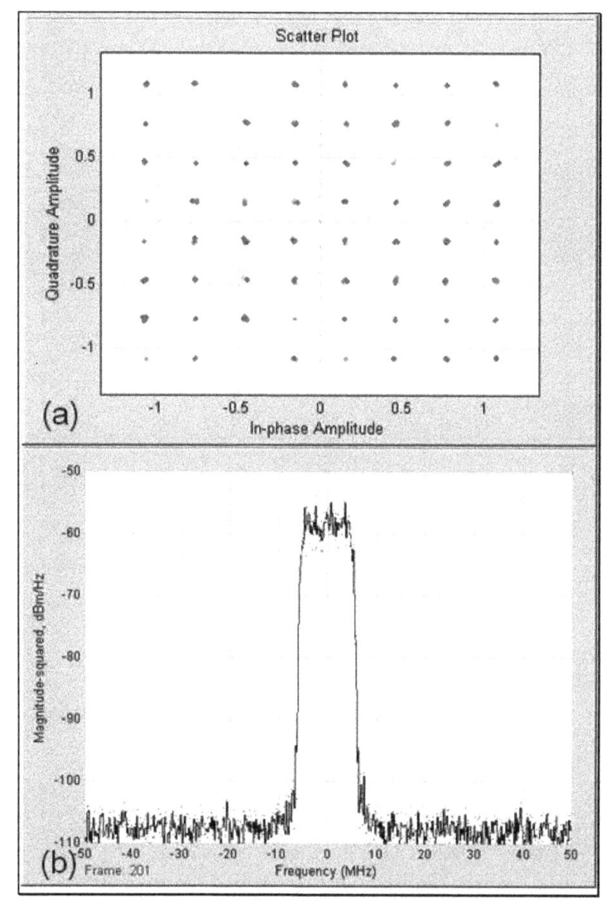

Figure 9. (**a**), Constellation and (**b**) spectrum of transmitted signal on all OAM modes (OAM1 + OAM8).

After modulation by the pseudo-Doppler waveforms, time-gating and down-conversion, the constellation and spectrum of the composite received signal appears as in Figure 10a,b respectively. The down-conversion frequency is 0 Hz in this case, but in general it will be some multiple of $2F$, where F is the pseudo-Doppler modulation frequency, because the gating pulses occur 2 times per cycle of F and they have harmonics.

Note that the spectrum of the receiver output composite signal consists of many harmonics of twice the pseudo-Doppler modulating frequency imposed at the receiver front end and is generally not symmetric about 0 Hz. Also note that the constellation does not look recoverable. The constellation is always obtained from the spectral replica positioned at 0 Hz after the down-converter.

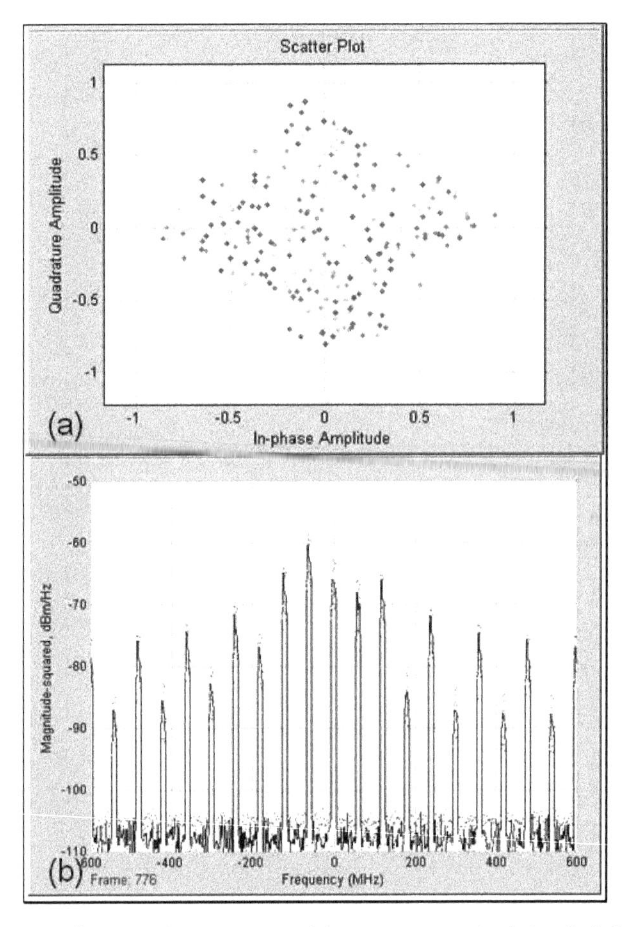

Figure 10. (**a**) Constellation, and (**b**) spectrum of the composite received signal of all OAM modes before demodulator (shifted by 0 Hz in this case).

It is interesting to observe the received constellation and spectrum of each OAM mode separately. This is shown, still with 0 Hz frequency shift in the down-converter, in Figure 11 below.

Note that, with a frequency shift of 0 Hz in the down-converter, OAM8 dominates over OAM1. With suitable scaling by a complex coefficient (amplitude and phase change), OAM8 could be recovered even in the presence of OAM 1, and after suitable conventional equalization and decoding, its QAM data symbols successfully demodulated.

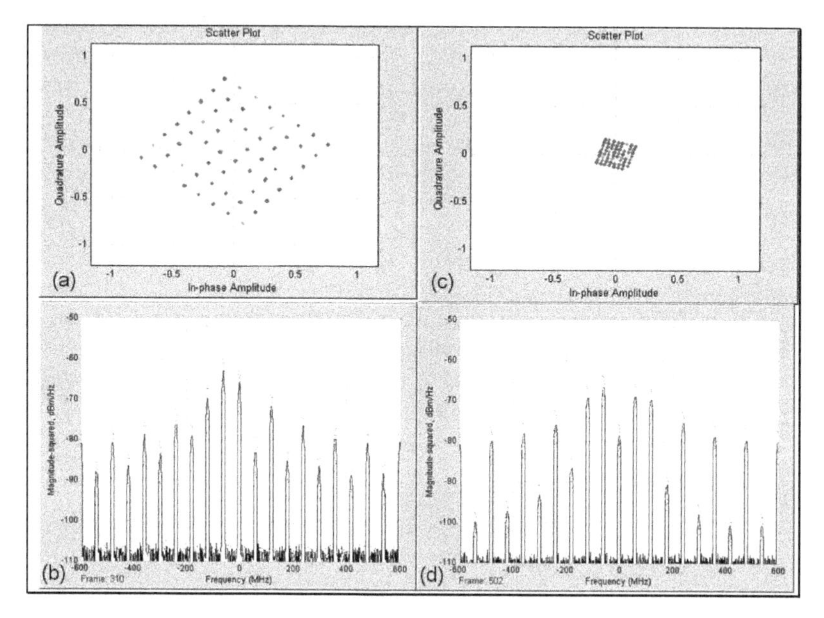

Figure 11. Received constellations and spectra of OAM modes transmitted separately and 0 Hz down-conversion: (**a**) constellation of OAM 8, (**b**) spectrum of OAM8, (**c**) constellation of OAM1, (**d**) spectrum of OAM1.

With a different frequency shift applied at the down-converter, other OAM modes can be recovered. For example, it turns out in this case that with a shift of $2F$, i.e., twice the pseudo-Doppler modulation frequency, the complementary situation arises, as shown in Figure 12.

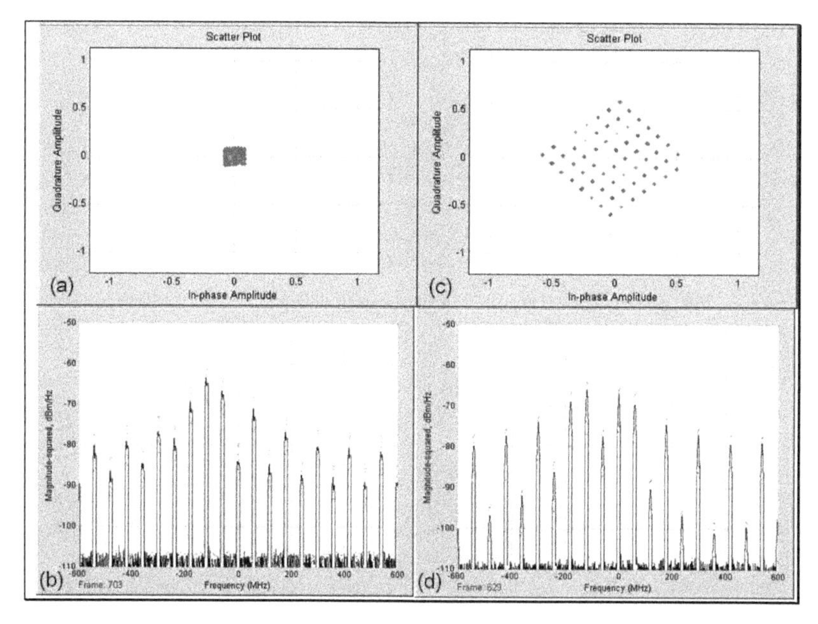

Figure 12. Separate OAM modes with shift by $2F$: (**a**) constellation of OAM8, (**b**) spectrum of OAM8, (**c**) constellation of OAM1, (**d**) spectrum of OAM1.

Now with frequency shift of $2F$ in the down-converter, OAM1 dominates over OAM8, so OAM1 could be similarly recovered and demodulated in the presence of OAM8. Without any other signal processing, each OAM mode was recovered from the composite signal with an uncoded error rate in the order of BER $\approx 10^{-1}$. These examples were chosen because the differences in OAM proportions happen to be very obvious, but this is not always the case. Moreover, the expected spectral shifts by fractions of the pseudo-Doppler modulation frequency appear to be absent in all of the output spectra.

Were the above simulation results a lucky coincidence and did they disprove the theory of the pseudo-Doppler frequency shifting of the OAM modes? It will be shown in the next subsection that this is not the case. There it is made clear that these results inspire a recovery algorithm for all the superposed OAM modes, and their spectral shifts will be shown to exist in the envelopes of the spectra. The differences in OAM proportions (more precisely, linear combinations) in the spectral replicas can be algebraically inverted so as to isolate them in final recovery outputs of a signal-processing subsystem. Such a subsystem would be based on least-mean-squares (LMS) optimization techniques and can be made adaptive to optimize their recovery in some statistical sense, much like existing MIMO receivers or adaptive-array systems.

4.3. OAM Recovery Algorithm

As evident from the above simulation experiments, the OAM signals are present in different proportions in the various harmonic spectral replicas (more accurately, "spectral shifts") of the composite received signal at the output of the gating subsystem. It is expected, and will indeed be shown, that these proportions are not random but fixed and deterministic, as are the relative amplitudes of the spectral replicas themselves (e.g., ranging from −60 to −90 dB in Figure 10). They are in fact determined by the physical parameters of the link, which can be made known to the receiver à-priori, thus enabling it to recover the OAM modes more effectively than was done in the simulation experiment. Specifically, several shifted spectra can be shifted to baseband and linearly combined so as to cause the amplitudes of the desired OAM mode to add and those of the undesired OAM modes to cancel coherently, using an LMS adaptive FIR (Finite Impulse Response) filter type of algorithm for each OAM mode. Subsequently, or as part of the LMS algorithm, the desired OAM mode is adjusted in amplitude and phase so its dynamic range matches that of the decision or demodulating subsystem, compensating for the dynamic range of the wireless link.

It is essential to recognize that the gating pseudo-Doppler modulations of the composite received signal constitute a "time-limited fractional frequency shift" operation on it in the discrete frequency domain. This can subsequently be recognized as the dual of a "frequency-limited fractional time shift", or band-limited fractional delay operation on a signal in discrete time domain as detailed in [19]. Specifically, the gating frequency, (which is twice the pseudo-Doppler modulation frequency), $2F$, and the fraction thereof comprising the OAM spectral shift, $kd/(2R)F$ as evident in (10), correspond to the sampling interval and fraction thereof, respectively, in the band-limited fractional-delay problem treated in [19]. This can be expected on the basis of the duality relations that exist between time and frequency domains due to properties of the Fourier transform and its inverse that relates them. The property that sampling in time-domain at intervals T causes periodic extensions in frequency-domain by $1/T$ also helps to explain the received spectra observed in the simulations.

Appendix C derives the frequency-domain effect of the time-limited fractional frequency-shift and its direct implementation along the same lines of reasoning as [19] for the impulse response and direct implementation of band-limited fractional delay in the time domain. The former can then be applied to (10) to plot the envelope of its spectrum and reveal the fractional pseudo-Doppler frequency shifts of the OAM modes. It also serves as the basis for an algorithm to recover the individual OAM modes from the gated output of the pseudo-Doppler modulator, as will be shown in the sequel.

In order to discern the spectral shift by the fractional pseudo-Doppler frequency, the Fourier transform of (10) is derived, in stages. First, (10) is affected by the time-gating and frequency-shift function so it is rewritten as a product of the cosine-modulated signal and the gating function with frequency-shift, inside the Fourier integral as

$$Z_{G,k}(f) = \sqrt{2}e^{j\psi_k} \int_{-\infty}^{\infty} S_k(t)\cos\left(\Omega t + \frac{\pi}{4}\right) g_k(t) e^{-j2\pi f t} dt \tag{14}$$

where $\psi_1(t) = 0$, $\psi_k = (1 - \pi/4)kd/(2R)$ were substituted. The time-gating function with frequency-shift, based on (13) is deduced to be the convolution

$$\begin{aligned} g_k(t) &= \int_{-\infty}^{\infty} u_k(\tau) \sum_{n=-\infty}^{\infty} (-1)^n \delta(t - \tfrac{n\pi}{2\pi F} - \tau) d\tau \\ with \quad & u_k(t) = e^{j2\pi F t(\frac{kd}{2R})} \; for \quad \tfrac{-\pi/2}{2\pi F} < t < 0 \\ and \quad & u_k(t) = 0 \quad otherwise \end{aligned} \tag{15}$$

The reasoning is that the gating function is a series of complex-valued pulses "shaped" as $u_k(t)$, repeating at intervals of π in Ωt (which is $1/(2F)$ in t), hence the convolution of $u_k(t)$ with the train of Dirac deltas. It is reasoned that the fractional pseudo-Doppler shift occurs only during the gating times, so those functions are coupled in the product. The additional feature is that the deltas alternate in sign. That feature may be absorbed by defining $g_k(t)$ as a product of the periodic extension of $u_k(t)$ with a square wave having period $1/F$ as in the simulation, where the phase-shift of $\pi/4$ centers the peaks in the gating intervals.

$$\Pi(t) = sign(-\sin(2\pi F t - \pi/4)) \tag{16}$$

and rewriting (15) more "cleanly" as

$$g_k(t) = \Pi(t) \int_{-\infty}^{\infty} u_k(\tau) \sum_{n=-\infty}^{\infty} \delta\left(t - \frac{n\pi}{2\pi F} - \tau\right) d\tau \tag{17}$$

with some foresight to the next stage of the derivation. That foresight is, that the above convolution has the Fourier transform $G_k(f)$ given by

$$G_k(f) = \Pi(f) * \left[U_k(f) \sum_{m=-\infty}^{\infty} \delta(f - m2F) \right] \tag{18}$$

where use was made of some identities involving Poisson sums, as explained in Appendix D, and $\Pi(f)$ is the Fourier transform of $\Pi(t)$. Before proceeding to evaluate (18), it is useful for later stages of this derivation, to express the integrand in (14) as the product of the transmitted signal $S_k(t)$ and the rest of the time function, calling it the pseudo-Doppler modulating receiver function $h_k(t)$, defined with the help of (17) as

$$h_k(t) = \cos\left(2\pi F t + \frac{\pi}{4}\right) g_k(t) \tag{19}$$

Now the gated output spectrum denoted by (14) can be expressed as the frequency-domain convolution of two Fourier transforms, namely (19) above convolved with $S_k(f)$, which is the Fourier transform of $S_k(t)$ and $2\pi F = \Omega$ as usual. Therefore, the gated output spectrum (14) can now be expressed as the convolution of (18) and (129):

$$Z_{G,k}(f) = H_k(f) * S_k(f) \tag{20}$$

A further simplification is afforded by combining the cosine in (19) with the sine in (16), reasoning that (16) can be adequately represented by its fundamental-frequency component and the signum function dispensed with, so (19) becomes

$$h_k(t) = -\cos\left(2\pi F t + \tfrac{\pi}{4}\right)\sin\left(2\pi F t - \tfrac{\pi}{4}\right) \times \cdots$$
$$\cdots \times \int_{-\infty}^{\infty} u_k(\tau) \sum_{n=-\infty}^{\infty} \delta\left(t - \tfrac{n\pi}{2\pi F} - \tau\right) d\tau \tag{21}$$

Using a trigonometric identity for the top line of (21) with $a/2 = 2\pi F t$ and $b/2 = \pi/4$ in

$$2\cos\left(\frac{a+b}{2}\right)\sin\left(\frac{a-b}{2}\right) = \sin(a) - \sin(b) \tag{22}$$

simplifies it to

$$h_k(t) = (-1/2)[\sin(2\pi 2Ft) - 1] \times \cdots$$
$$\cdots \times \int_{-\infty}^{\infty} u_k(\tau) \sum_{n=-\infty}^{\infty} \delta\left(t - \tfrac{n\pi}{2\pi F} - \tau\right) d\tau \tag{23}$$

Now, with the help of (23), the final output spectrum (30) can be expressed as

$$\boxed{Z_{G,k}(f) = \Pi_{CS}(f) * \left[U_k(f) \sum_{m=-\infty}^{\infty} S_k(f - m2F)\right]} \tag{24}$$

with the understanding that

$$\Pi_{CS}(f) = \left(\tfrac{-1}{2}\right) \int_{-\infty}^{\infty} [\sin(2\pi 2Ft) - 1] e^{-j2\pi f t} dt = \cdots$$
$$\cdots = \left(\tfrac{1}{2}\right) \int_{-\infty}^{\infty} \left[e^{-j2\pi f t} - \tfrac{e^{j-2\pi(f-2F)t} - e^{-j2\pi(f+2F)t}}{j2}\right] dt = \cdots$$
$$\cdots = \delta(f)/2 + j\delta(f - 2F)/4 - j\delta(f + 2F)/4 \tag{25}$$

This means that the spectrum of the gated output is a periodic extension of the transmitted spectrum of the signal with repetition interval equal to twice the pseudo-Doppler modulation frequency, $2F$, multiplied by the spectral envelope $U_k(f)$.

The next step is to evaluate $U_k(f)$, which is the envelope of the spectrum, and manifests the fractional pseudo-Doppler frequency shift expected according to the order, k, of the OAM mode. This shift is traceable to the complex "pulse shape" function $u_k(t)$ defined in (15). Then the convolution with (25) is performed at the end. The spectrum of the envelope is

$$U_k(f) = \int_{\frac{-\pi/2}{2\pi F}}^{0} e^{j2\pi\left(\frac{kd}{2R}\right)Ft} e^{-j2\pi f t} dt = \int_{\frac{-1}{4F}}^{0} e^{-j2\pi(f-\mu_k)t} dt \tag{26}$$

where the fractional pseudo-Doppler shift is $\mu_k = F(kd/(2R))$ in accordance with Appendix C. It is straight-forward but tedious to evaluate, producing

$$U_k(f) = \frac{e^{j\frac{\pi(f-\mu_k)}{4F}}}{4F}\left[\frac{\sin\left(\frac{\pi(f-\mu_k)}{4F}\right)}{\left(\frac{\pi(f-\mu_k)}{4F}\right)}\right] \tag{27}$$

Then according to (24), the convolution of (27) with (25) gives the complete spectral envelope

$$U_{k,CS}(f) = \Pi_{CS}(f) * U_k(f) = \cdots$$
$$\cdots = \tfrac{1}{2}U_k(f) + \tfrac{j}{4}[U_k(f - 2F) - U_k(f + 2F)] \tag{28}$$

Therefore, a sufficiently representative expression for the final output spectrum can be obtained by using (28) in (24) to obtain the output spectrum

$$Z_{G,k,1}(f) = U_{k,CS}(f) \sum_{m=-\infty}^{\infty} S_k(f - m2F) \tag{29}$$

It remains to evaluate (28) and plot its amplitude, so as to visualize the fractional pseudo-Doppler frequency shift at the output of this receiving subsystem.

The magnitude of (28) is plotted in the subsequent Figure for the parameters used in the simulation with k = 8 in $\mu_k = F(kd/(2R))$. What is actually plotted is the amplitude

$$\begin{aligned} E_k(f) &= \left(\tfrac{1}{2}\right) Sinc\left(\tfrac{f-\mu_k}{4F}\right) + \cdots \\ \cdots &+ \left(\tfrac{1}{4}\right) Sinc\left(\tfrac{f-2F-\mu_k}{4F}\right) + \cdots \\ \cdots &+ \left(\tfrac{1}{4}\right) Sinc\left(\tfrac{f+2F-\mu_k}{4F}\right) \end{aligned} \tag{30}$$

with the common factor

$$P_k(f) = \frac{1}{4F} e^{j\frac{\pi(f-\mu_k)}{4F}} \tag{31}$$

being omitted for clarity. The spectral replica according to (29) are superposed at intervals of $2F$.

Equation (29) is representative of the final output of the gating subsystem of the pseudo-Doppler modulated OAM receiver. Clearly, the spectral replica are spaced at twice the pseudo-Doppler frequency, $2F$, while the peak of the envelope is at the fractional pseudo-Doppler frequency, $\mu_k = F(kd/(2R))$, as expected and as observed in the simulation results. More accurate representation of the spectral envelope can be obtained by including higher harmonics of F when simplifying (16).

Now that it has been established that the gated output (29) consists of different proportions of spectral replica of each OAM mode, an algorithm for recovering them may be proposed. First it will be necessary to truncate the series of spectral replicas to a minimum equal to the number of OAM modes to be recovered, because at least that many different linear combinations of them will be needed. Each linear combination of spectral replicas is determined by the spectral envelope with its unique fractional pseudo-Doppler shift as in Figure 13. An arrangement such as in Figure A6 can be used to combine all the replicas in such a way that only those of the desired OAM mode will add up to a non-zero complex amplitude while those of all other OAM modes will cancel to zero, much like in adaptive-array signal-processing which nulls interfering sources' signals. The coefficients should be low-pass so only the baseband spectral replicas are passed, as in down-conversion.

The OAM recovery process can be understood in terms of using (29) truncated to $M = K$ terms and transformed into time-domain, as $x(t)$ in (A14), where K is at least the number of OAM modes. The coefficients of the down-converted signals at each stage will be different than in (A14); they will now be derived jointly for all K of the OAM modes. Note that each stage has as input to its coefficient one of the spectral replicas (index "m" in (29)) of all the OAM modes superposed with their amplitudes as received upon. The stage inputs to the coefficients can be arranged in an $Mx1$ vector whose m-th row entry is

$$x_m(t) = \sum_{k=1}^{K} U_{k,m} \underbrace{a_k s_k(t) e^{-j2\pi(m2F)t}}_{Lowpass} \tag{32}$$

and $U_{k,m}$ is the spectral envelope coefficient for the k-th OAM mode at the m-th spectral replica (i.e., stage coefficient). The lowpass signal at each stage is the same (because the spectral replicas are in fact replicas of the same signal, composed of all the OAM modes in their arrival proportions $\{a_k\}$) so in vector-matrix form, (32) can be written as

$$\mathbf{X} = \sum_{k=1}^{K} U_k a_k s_k = \cdots$$
$$\cdots = \begin{bmatrix} \mathbf{U}_1 & \mathbf{U}_2 & \cdots & \mathbf{U}_K \end{bmatrix} \begin{bmatrix} s_1 & & & 0 \\ & s_2 & & \\ & & \ddots & \\ 0 & & & s_K \end{bmatrix} \begin{bmatrix} a_1 \\ a_2 \\ \vdots \\ a_K \end{bmatrix} \tag{33}$$

where the time arguments were omitted for clarity. Further compacting the notation, one can write (33) as

$$\mathbf{X}_{M\times 1} = \mathbf{U}_{M\times K}\mathbf{S}_{K\times K}\mathbf{A}_{K\times 1} \tag{34}$$

where the matrix dimensions are shown explicitly to help with the defining correspondence with (33), with $M \geq K$.

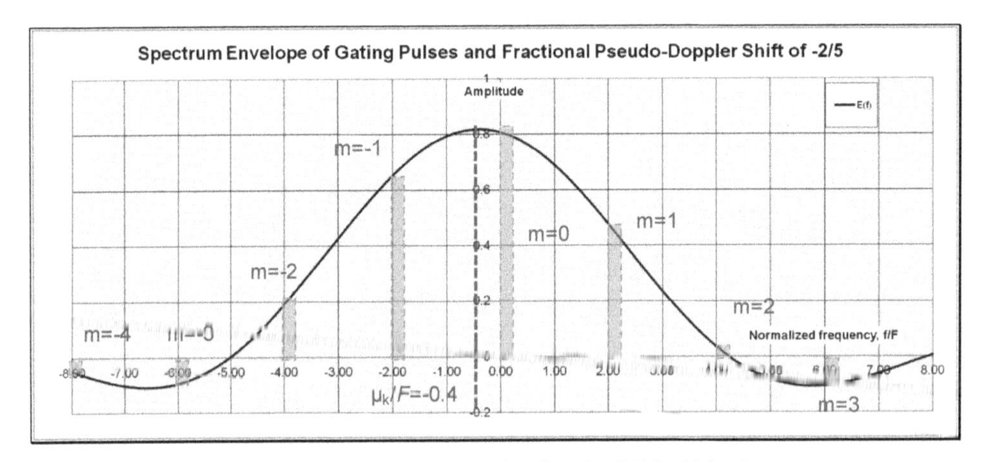

Figure 13. Spectral envelope and replicas for OAM with k = 8.

In order to recover all K of the OAM modes jointly, one needs K branches of the kind shown in Figure A6, each with M coefficients. That amounts to a KxM coefficient matrix $\mathbf{C}$, which is derived next. When vector $\mathbf{X}$ is pre-multiplied by $\mathbf{C}$, it will produce an output vector $\mathbf{Y}$ whose entries are the separated OAM mode signals. In fact, the OAM modes are present in their arrival proportions (which is understandable because the algorithm has no information about their transmitted proportions), so they will need to be equalized before they can be demodulated in the "conventional" way. The output vector is written as

$$\mathbf{Y}_{K\times 1} = \mathbf{C}_{K\times M}\mathbf{X}_{M\times 1} = \mathbf{S}_{K\times K}\mathbf{A}_{K\times 1} \tag{35}$$

By inspection of (34), it looks like the RHS of (35) may be obtained by pre-multiplying vector $\mathbf{X}$ by the inverse of matrix $\mathbf{U}$. However, matrix $\mathbf{U}$ is not always square (unless $M = K$), so its inverse does not exist. Fortunately, the dimensions of the matrices and vectors involved are such that a pseudo-inverse does exist, which then becomes the desired coefficient matrix by correspondence with (35) as follows: Pre-multiply both sides of (34) by the Hermitian (complex-conjugate transpose) of matrix $\mathbf{U}$, denoted as $\mathbf{U}^H$ to obtain

$$\mathbf{U}^H_{K\times M}\mathbf{X}_{M\times 1} = \mathbf{U}^H_{K\times M}\mathbf{U}_{M\times K}\mathbf{S}_{K\times K}\mathbf{A}_{K\times 1} \tag{36}$$

Now notice that $\mathbf{U}^H\mathbf{U}$ is a $K\times K$ square matrix, so it can be invertible and one can pre-multiply both sides of (36) by it.

$$\begin{aligned}
&\left[\mathbf{U}_{K\times M}^{H}\mathbf{U}_{M\times K}\right]^{-1}\mathbf{U}_{K\times M}^{H}\mathbf{X}_{M\times 1} = \cdots \\
&\cdots = \left[\mathbf{U}_{K\times M}^{H}\mathbf{U}_{M\times K}\right]^{-1}\mathbf{U}_{K\times M}^{H}\mathbf{U}_{M\times K}\mathbf{S}_{K\times K}\mathbf{A}_{K\times 1} = \cdots \\
&\cdots = \mathbf{S}_{K\times K}\mathbf{A}_{K\times 1}
\end{aligned} \quad (37)$$

Denoting the pseudo-inverse of matrix **U**, which is found on the LHS of (37), as $\mathbf{U}^{\#}$ i.e.,

$$\mathbf{U}_{K\times M}^{\#} = \left[\mathbf{U}_{K\times M}^{H}\mathbf{U}_{M\times K}\right]^{-1}\mathbf{U}_{K\times M}^{H} \quad (38)$$

allows one to write (37) as is commonly done in least-squares optimization problems

$$\mathbf{U}_{K\times M}^{\#}\mathbf{X}_{M\times 1} = \mathbf{S}_{K\times K}\mathbf{A}_{K\times 1} \quad (39)$$

so by correspondence with (35), the joint OAM recovery coefficients matrix is

$$\mathbf{C}_{K\times M} = \mathbf{U}_{K\times M}^{\#} \quad (40)$$

Therefore, the recovered OAM modes are obtained from the input vector **X** simply as

$$\mathbf{Y}_{K\times 1} = \mathbf{U}_{K\times M}^{\#}\mathbf{X}_{M\times 1} = \mathbf{S}_{K\times K}\mathbf{A}_{K\times 1} \quad (41)$$

which can be subsequently equalized and demodulated as in the "back end" (or DSP baseband section) of a "conventional" MIMO digital radio receiver. Although the matrix inverse found within the pseudo-inverse in (38) is not guaranteed to exist, it is more likely the more spectral replica are included, i.e., the larger the M is. It is dependent on the differences among the spectral envelopes of each of the OAM modes, which become more apparent as more spectral replica are included with increasing M. Increasing M beyond K does not increase the dimensions of $\mathbf{U}^{H}\mathbf{U}$, but can improve its condition number, thus enhancing its invertibility. In other words, including more spectral components beyond the minimum number "K" can lead to better least-squares estimates of the OAM signals, with smaller error-vector magnitudes (EVMs) due to crosstalk.

Note also that this OAM recovery algorithm is deterministic because all the information contained in matrix **U** is known at the receiver, in principle. (The distance R from the beam axis may be deduced in non-fixed link via other signaling information such as timing-advance in TDD systems and from the TX antenna array geometry.) In practice **U** could be obtained by measuring the spectral-envelope coefficients of each OAM mode using correlation techniques during periodic "calibration" intervals, when each OAM mode would be transmitted separately. The coefficients for each branch according to (40) are shown explicitly in a sketch of the recovery algorithm in Figure 14 below (based on Figure A6 on Appendix C).

With suitable training signals for reference, an LMS type of algorithm can be formulated using well-known feedback loops to adapt the coefficients in the FIR-type of filter structure, as mentioned earlier. Moreover, including more than K spectral components and corresponding coefficient loops would provide more degrees of freedom, which could lead to further improved EVMs and potentially cancel external interference.

The structure in Figure 14 is reminiscent of a spectral-analysis process, so it may be feasible to implement equivalent versions of it using Fourier transform techniques. An acousto-optic spectrum-analyzer configuration comes to mind and may be pursued in future work.

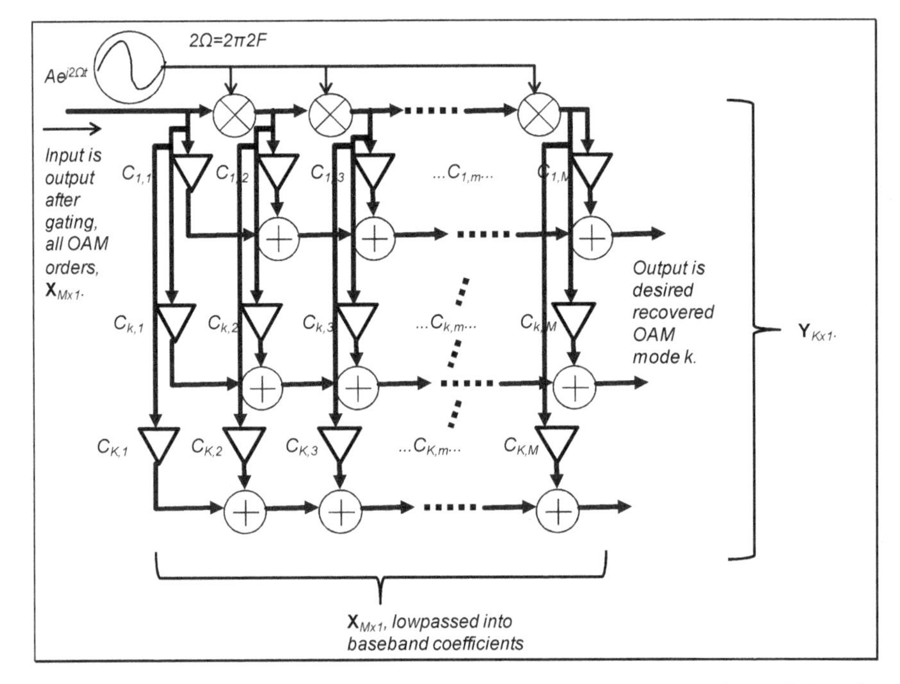

Figure 14. Functional OAM recovery algorithm from gated output of pseudo-Doppler modulation subsystem.

It is also possible to make use of output Z_2, noting that the output at Z_2 can be derived by the same process as was that at Z_1 but starting with

$$Z_{2,k}(t) = S_k(t)e^{j\psi_1(t)}\left[\sin(\Omega t) + e^{\frac{-jkd}{R}}\cos(\Omega t)\right] \tag{42}$$

One obtains the corresponding result for Z_2 prior to the gating operation, as

$$\begin{aligned} Z_{2,k}(t) &\cong \sqrt{2}S_k(t)e^{j(\psi_1(t)-\frac{kd}{2R})} \times \cdots \\ &\cdots \times (\cos(P(t) - \pi/4))e^{j\tan^{-1}((\frac{kd}{2R})\tan(P(t)-\pi/4))} \end{aligned} \tag{43}$$

which looks like $\pi/4$ was replaced by $-\pi/4$ in $Z_{1,k}(t)$. Although one can make (10) and (43) look like the input by substituting $\cos(\Omega t - \pi/4) = \cos(\Omega t + \pi/4 - \pi/2) = \sin(\Omega t + \pi/4)$ in (43), the change from $\pi/4$ to $-\pi/4$ also forces a change in the gating intervals. Using the same procedure as in Appendix D, it is found that (43) needs to be gated to intervals where

$$0 < \Omega t < \frac{\pi}{2} \tag{44}$$

also separated by multiples of π, which are exactly complementary to those shown in Figure 8. That means the useful outputs at Z_1 and Z_2 after their respective gating operations appear in disjoint time intervals, so it makes no sense to combine them even though they have the same fractional pseudo-Doppler shifts that are desired. What does make sense is to "toggle" them to the same final output in order to have a more time-continuous output signal for the OAM recovery algorithm leading to (41), which may become useful in future refinements.

5. Conclusions

In this paper, a technique for recovering signals from received OAM beams based on pseudo-Doppler effect in the frequency domain was developed. Whereas reference [10] developed a method for detecting individual OAM modes based on an interpolation technique involving a minimum of two receiving antennas positioned tangentially on the peak region of OAM beams in the far field, this paper advanced the method to the point of recovering information carried on multiple OAM beams simultaneously, amenable to real-time implementation. Preliminary simulation results confirmed the ability of this technique to recover the modulation signals from several OAM beams in accordance with the mathematical least-squares type of signal-processing carried out in the frequency domain, based on the effect that the pseudo-Doppler technique imparts a different fractional pseudo-Doppler frequency shift to each received OAM mode. This technique was motivated by the observations that current spatial techniques for receiving and recovering the OAM modes in unguided radio-frequency applications are limited by the conical shapes of the (non-zero-order) OAM beams, which result in poor SNR, impractically large receiving antennas, short link distances, applicability to only very short wavelengths or guided propagation, or a combination thereof. These limitations are a consequence of attempting to receive and recover the OAM modes by the reverse of the spatial technique of transmitting them using co-axially situated circular antennas, motivated by increasing spectral efficiency of wireless links via spatial multiplexing. Here the spatial multiplexing applies at the transmitter, but the receiving technique essentially transforms the recovery problem into the frequency domain inside the receiver, keeping the radiated bandwidth the same. Other efforts at increasing data throughput of wireless links are directed at exploiting wide spectral bandwidth where available, using ultra wide-band (UWB) techniques, specifically UWB antennas [20]. Such efforts could be combined with the present OAM technique to further increase transmission capacity in the future.

Funding: This research received no external funding.

Acknowledgments: The author wishes to thank Huawei Canada R&D Center, Ottawa, for the support of this work-in-progress.

Conflicts of Interest: The authors declare no conflicts of interest.

Appendix A. The Doppler Shift

For completeness, the shift in the frequency of a propagating wave due to relative motion between source and receiver is reviewed.

Consider a stationary point at $x = 0$ with a wave incident on it at its propagation velocity, c, along the direction of the x-axis. Define the phase of the wave to be 0 there. With the help of Figure A1, the phase of any other point $x(t)$ along the x-axis at the same instant in time is given by

$$\psi(x) = 2\pi\left(\frac{x(t)}{\lambda}\right) \tag{A1}$$

Next, let the point $x(t)$ move along the x-axis at a constant velocity v. Because the phase is linear with position, this will cause a corresponding linear change in its phase as it moves, given by

$$\frac{d}{dt}\psi(x(t)) = \frac{d\psi(x(t))}{dx}\frac{dx(t)}{dt} = \left(\frac{2\pi}{\lambda}\right)\frac{dx(t)}{dt} \tag{A2}$$

The rate of change of phase with respect to time is defined as radian frequency, $2\pi F_d$, while the rate of change of distance with time is simply velocity, v, therefore

$$\frac{d}{dt}\psi(x(t)) = 2\pi F_d = \left(\frac{2\pi}{\lambda}\right)v \tag{A3}$$

or simply

$$F_d = \left(\frac{v}{\lambda}\right) = f\left(\frac{v}{c}\right) \tag{A4}$$

where F_d is recognized as the (axial) Doppler frequency and f is the frequency of the incident wave. Although the incident wave is moving at the same time that point $x(t)$ is moving, all this is calculated at a single point in time, as if time was frozen, so the position and velocity of the point x and the incident wave are assumed known at the instant t, Heisenberg notwithstanding. One can simply visualize point $x(t)$ moving along the frozen wave, at each position reading its corresponding phase; the frozen wavelength constitutes a phase gradient from 0 to 2π radians.

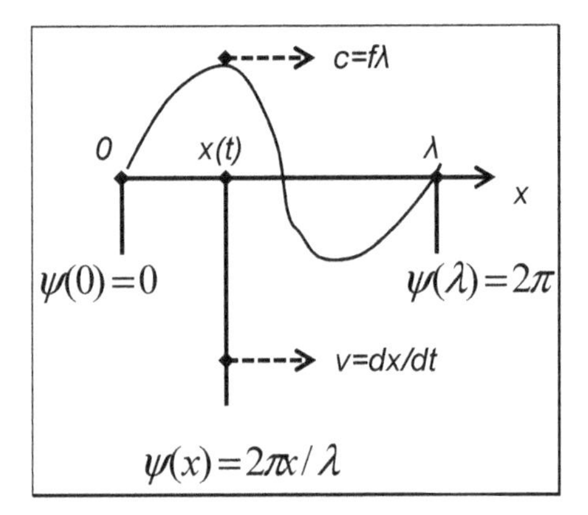

Figure A1. Derivation of axial Doppler shift.

When a wave with OAM is incident on a point, the received phase can experience a frequency shift even as it moves in a direction transverse to the direction of propagation. Specifically, the observation point here will be moving in the direction of θ along the circumference of the peak of the conical OAM beam in Figure 7 of the main text, where a phase gradient also exists at a given point in time. It is shown below as the "unwrapped" footprint of Figure 4 of the main text.

Again, the electrical phase at the observation point $R\theta = d$ is d times the phase gradient k/R, equal to the proportion that d constitutes of the circumference $2\pi R$, times k cycles of 2π that the k-th order OAM beam imposes on that circumference of its footprint. It will vary in time if its position is changing in time, because it is passing through the phase gradient, so

$$\frac{d}{dt}\psi(R\theta(t)) = \frac{d\psi(R\theta(t))}{d\theta}\frac{d\theta(t)}{dt} = \left(\frac{2\pi k}{2\pi R}\right)R\frac{d\theta(t)}{dt}$$

Simplifying in terms of the physical variables, the rate of change of electrical phase with time again constituting an angular frequency,

$$\frac{d}{dt}\psi(R\theta(t)) = 2\pi\left(\frac{kR}{2\pi R}\right)\frac{d\theta(t)}{dt} = k\frac{d\theta(t)}{dt} = 2\pi F_T$$

Recognizing that $v = Rd\theta/dt$, the above reduces to

$$2\pi F_T = \frac{kv}{R} \tag{A5}$$

where F_T is the transverse Doppler frequency in Hz. Note that there is now no dependence on RF carrier frequency, f, of the OAM beam. Note that this is not the transverse Doppler shift that is sometimes derived in physics context using relativity arguments.

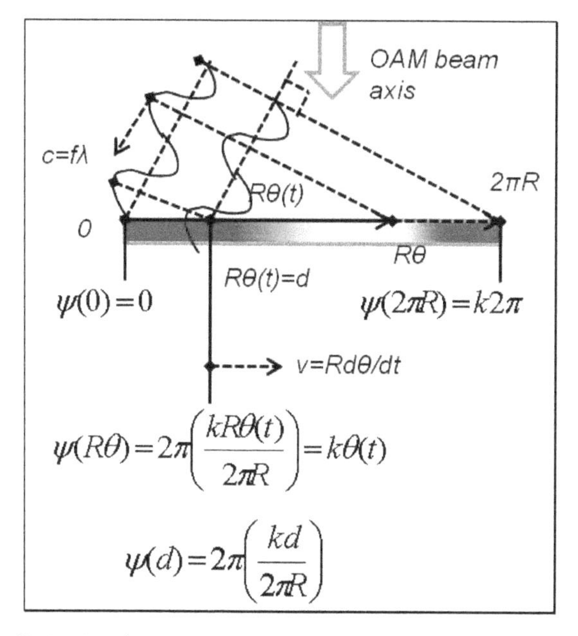

Figure A2. Derivation of transverse-type, or rotational Doppler effect in an OAM beam.

One may wonder why in Figure A2, the wave fronts are not propagating in the same direction as the beam axis but are angled relative to it. (The angle is shown exaggerated.) The reason is that the beam has OAM, which causes its Poynting vector to follow a helical "corkscrew" path, orthogonal to the wave fronts which form helical paths around the beam axis.

For an OAM beam of order k, one can visualize the phase fronts (wave fronts) as k parallel threads wrapped around the beam axis like the threads of a machine screw. The Poynting vector would then be orthogonal to them everywhere in the beam. The higher the OAM order k, the larger is the angle the k parallel phase threads make with the beam axis (and with its transverse footprint, where its phase gradient becomes steeper in proportion to k). That is why the transverse Doppler frequency is directly proportional to OAM order k.

It is this motion (also expressed in (5) of the main text) of the receiving antenna through this OAM phase gradient at the footprint, which the pseudo-Doppler technique attempts to emulate. It is necessary to emulate a very high transverse Doppler frequency F_T, so that each OAM mode can be recovered without its spectrum overlapping those of the neighboring OAM modes when the modulated carrier wave has a high bandwidth, B.

Appendix B. Bessel Functions of the First Kind, $J_k(x,)$ Plotted

Here the Bessel functions of the first kind are plotted for the first few integer orders.

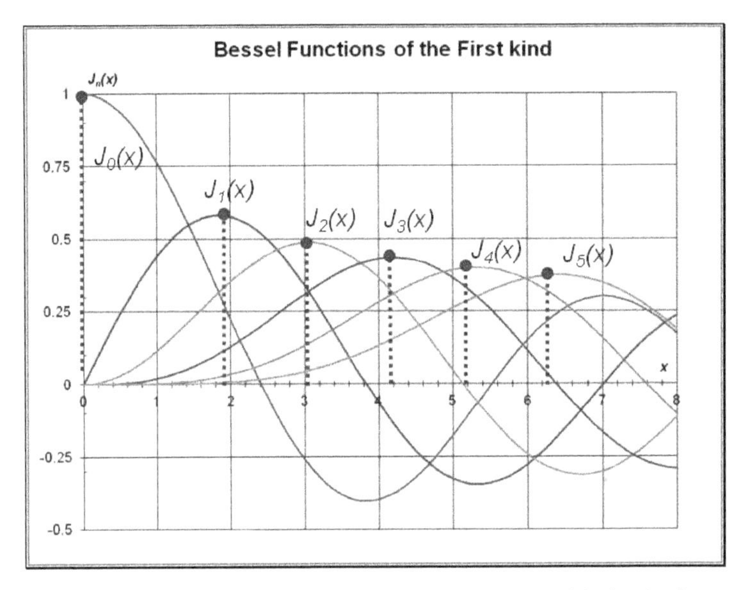

Figure A3. Plots of the first few orders of Bessel functions of the first kind.

Appendix C. Development of the Time-Limited Fractional Frequency Shift

In this appendix, the discrete frequency-domain characterization of the "time-limited fractional frequency shift" operation will be developed following the principles of its dual in discrete time-domain, namely the "band-limited fractional time delay" operation according to [19].

What makes it possible to synthesize a delay, $\tau = pT$, which is only a fraction of the sample interval, T, in a discrete-time system sampled at rate $1/T$, is the fact that this fractional delay needs to be effective only in a limited bandwidth around 0 Hz, at most as wide as the Nyquist band, $B = 1/T$. Therefore, from $f = -1/(2T)$ to $1/(2T)$, the frequency response $H(f)$ should be all-pass, that is flat with phase factor $e^{-j2\pi f\tau}$, and zero outside this band, making it low-pass. The corresponding impulse response in continuous-time is

$$h(t) = \int_{-\infty}^{\infty} H(f)e^{j2\pi ft}df = \int_{-B/2}^{B/2} e^{j2\pi f(t-\tau)}df \tag{A6}$$

which is straight-forward to evaluate as

$$h(t) = B\frac{\sin(\pi B(t-\tau))}{\pi B(t-\tau)} \equiv BSinc(B(t-\tau)) \tag{A7}$$

This defines the envelope of the discrete-time samples taken at integer multiples of T by the sampling impulses as

$$h(nT) = \int_{-\infty}^{\infty} h(t)\delta(t-nT)dt = BSinc(B(nT-\tau)) \tag{A8}$$

The envelopes and samples of the impulse response are shown below for both integer and fractional delays, computed according to Figure 3 of [19].

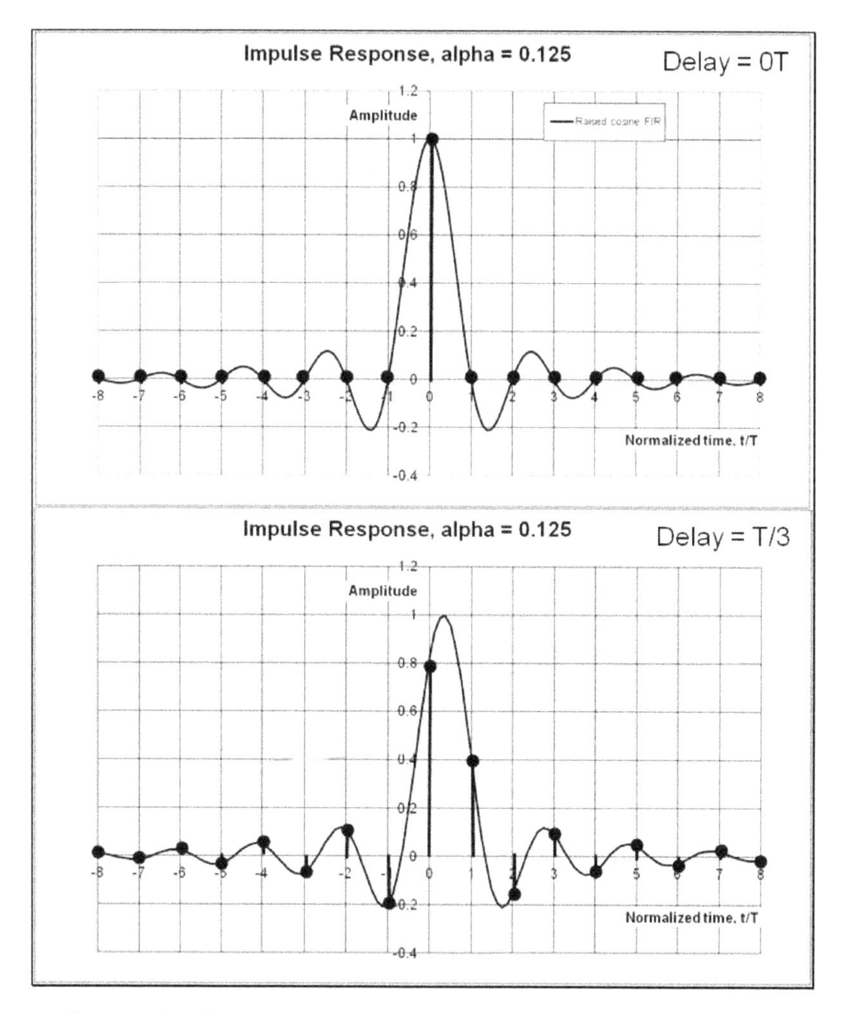

Figure A4. Impulse responses of band-limited integer and fractional delay filters [19].

By substituting $T = 1/B$ and $\tau = pT$, this becomes the discrete-time impulse response

$$y(N) = B\sum_{n=0}^{N} x(N-n)Sinc(n-p) \tag{A9}$$

which can be implemented directly as a FIR filter whose coefficients correspond to integer values of n with $0 < p < 1$ in the above summation. In practice the filter s truncated to a practical number of coefficients, $N + 1$, corresponding to N sample delays, and effects a total delay of $NT/2 + \tau$, because it must be causal. It is reproduced below according to Figure 4 of [19].

It can also be utilized in an adaptive multipath equalizer by adapting the coefficients according to a feedback algorithm that minimizes some statistical property of the signal error, as for example in [19].

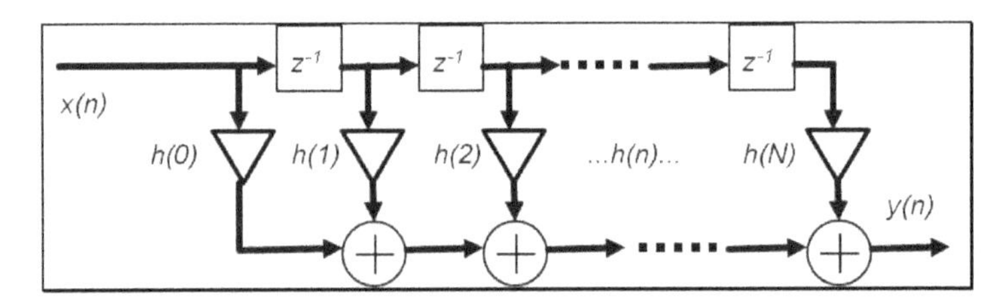

Figure A5. Direct realization of band-limited fractional-delay filter [19].

A parallel development of the dual structure for the fractional frequency shift of $\mu = q2F$ in frequency domain can now proceed as follows: denoting the time-gating function as $g(t) = e^{j2\pi\mu t}$ for $-A/2 < t < A/2$ and 0 elsewhere, its spectrum in the continuous frequency-domain is the inverse Fourier transform

$$G(f) = \int_{-\infty}^{\infty} g(t)e^{-j2\pi ft}dt = \int_{-A/2}^{A/2} e^{-j2\pi(f-\mu)t}dt \tag{A10}$$

which is seen to evaluate to

$$G(f) = A\frac{\sin(\pi A(f-\mu))}{\pi A(t-\mu)} \equiv ASinc(A(f-\mu)) \tag{A11}$$

This defines the envelope of the frequency-response of the fractional-shift operation, where it is clearly seen that its peak is shifted from the center 0 Hz to $f = \mu$ Hz. When this gating window is repeated in time at intervals $T/2$ (periodic extension in time), it results in spectral sampling at intervals $2F - 2/T$ expressed as

$$\begin{aligned} G(m2F) &= \int_{\infty} G(f)\delta(f - m2F)df \cdots \\ \cdots &= ASinc(A(m2F-\mu)) \end{aligned} \tag{A12}$$

finally resulting in the discrete-frequency response

$$Y(M) = \frac{1}{2F}\sum_{m=0}^{M} X(m-M)Sinc(m-q) \tag{A13}$$

with $A = 1/(2F)$ and $\mu = q2F$. It is the dual of (C4) and defines the coefficients of the spectral replicas of the frequency-response of the time-gating function. A conceptual realization analogous to Figure A5 can be formulated for it, truncated to the M-th harmonic of $2F$. Because the realization must be in (real-) time domain, the "$2F$" frequency-shift elements corresponding (in dual fashion) to the time-shift (sample delay) elements z^{-1} in Figure A5 are replaced by multipliers by complex factor $e^{j2\pi 2Ft}$, which effects the harmonics of 2 times the pseudo-Doppler modulating frequency, $2F$. Accordingly, in time-domain the input–output relation in continuous time becomes

$$y(t) = \frac{1}{2F}\sum_{m=0}^{M} x(t)e^{j2\pi(m-M)2Ft}Sinc(m-q) \tag{A14}$$

In this application, the time gating intervals are not symmetric about 0 and they alternate in sign at every half period of the modulating frequency, F, so the spectrum envelopes will be different, but the same fractional pseudo-Doppler shifts, $Fkd/(2R)$, will be observed in their envelopes. These shifts are not directly visible in the simulation results because the spectrum envelopes are not plotted, only the spectral replicas at the repetition frequencies according to (A11), with the correspondence of $q = kd/(4R)$ in $\mu = q2F = Fkd/(2R)$.

As in the case of the fractional sample delay filter, a similar structure based on (A13) can be used in the "equalization" or recovery of OAM modes in the frequency domain (but still has to be implemented in time domain as discussed above). In this application, the recovery algorithm can be more deterministic because the necessary parameters (those defining the fractional shift q above) can be made known to the receiver.

For reference and to complete the duality statement, the following structure is formulated as a basis for the OAM recovery algorithm.

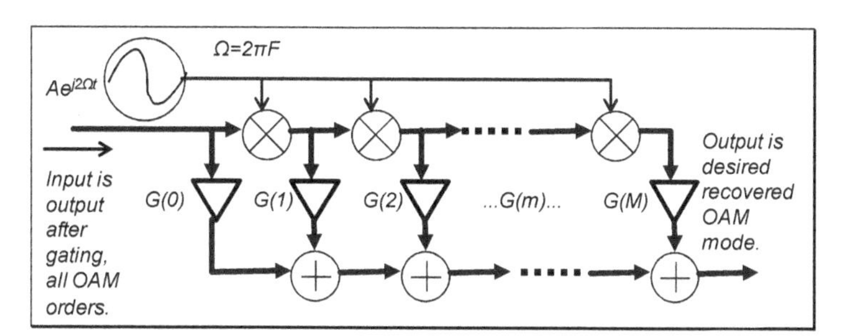

Figure A6. Direct representation of time-limited frequency-shift structure used for recovery of OAM mode.

The structure in Figure A6 is a functional representation of (A14) truncated to $M + 1$ terms and can be used to recover a desired OAM mode from a superposition of all OAM modes that appears at the output of the gating block of the pseudo-Doppler modulation subsystem of the receiver front end. Note that one set of coefficients $\{G(m)\}$ is required to recover each order of OAM mode.

It is important to remember that the entire broad-band spectrum containing M harmonics of the pseudo-Doppler modulation, $2F$, and the associated spectral replicas of the transmitted signal. Note also that the input and output signals in Figure A6 are still in continuous-time domain, and may also be in analog form.

Appendix D. Some Identities Involving Poisson Sums

Here it is derived that

$$\int_{-\infty}^{\infty} e^{-j2\pi ft} \sum_{n=-\infty}^{\infty} \delta(t - nT)dt = \frac{1}{T} \sum_{m=-\infty}^{\infty} \delta\left(f - \frac{m}{T}\right) \tag{A15}$$

Alternative derivations can be found at https://en.wikipedia.org/wiki/Poisson_summation_formula.

Before proceeding, it is worth keeping in mind that multiplying a time function by a string of Dirac deltas as found in the LHS of (A15) constituters sampling it, which corresponds to convolving its Fourier spectrum by the string of Dirac deltas on the RHS of (A15), which constitutes a periodic extension of its Fourier spectrum. That is effectively what was done in the main text in relation to (14)–(21), but the sampling function was not just Dirac deltas, but pulses with a complex-valued "shape", which is a convolution of that pulse shape with a string of Dirac deltas as expressed by (17).

So on the LHS is the Fourier transform of a string of Dirac deltas, which is a periodic function of time. As such, this periodic function has a Fourier series with coefficients $\{c_m\}$ and can be expressed as a sum of the harmonics of the fundamental repetition frequency, F, which is the reciprocal of the repetition period, T.

$$\sum_{n=-\infty}^{\infty} \delta(t - nT) = \sum_{m=-\infty}^{\infty} c_m e^{j2\pi mFt} \tag{A16}$$

The Fourier coefficients are determined by the usual dot product of the periodic function with the complex conjugate of harmonic basis function, i.e., the integral over one period, normalized by the period length

$$c_m = \frac{1}{T}\int_0^T \sum_{n=-\infty}^{\infty} \delta(t-nT)e^{-j2\pi\frac{m}{T}t}dt \tag{A17}$$

Only the $n = 0$ period of the string of Dirac deltas falls within the limits of the integral, so (A17) becomes

$$c_m = \frac{1}{T}\int_0^T \delta(t)e^{-j2\pi\frac{m}{T}t}dt = \frac{1}{T}e^{-j2\pi\frac{m}{T}0} = \frac{1}{T} \tag{A18}$$

which means the coefficients are all equal to $1/T$. That simplifies (A16) to

$$\sum_{n=-\infty}^{\infty} \delta(t-nT) = \frac{1}{T}\sum_{m=-\infty}^{\infty} e^{j2\pi mFt} \tag{A19}$$

That is now substituted in the LHS of (A15) and the order of summation and integration is reversed, as each operation is linear, and really a form of addition.

$$\int_{-\infty}^{\infty} e^{-j2\pi ft}\frac{1}{T}\sum_{m=-\infty}^{\infty} e^{j2\pi mFt}dt = \frac{1}{T}\sum_{m=-\infty}^{\infty}\int_{-\infty}^{\infty} e^{-j2\pi(f-mF)t}dt \tag{A20}$$

The integral is now evaluated as a limit:

$$\begin{aligned} \int_{-\infty}^{\infty} e^{-j2\pi(f-mF)t}dt &= \lim_{\tau\to\infty}\int_{-\tau/2}^{\tau/2} e^{-j2\pi(f-mF)t}dt = \cdots \\ \cdots &= \lim_{\tau\to\infty}\left[\frac{e^{-j2\pi(f-mF)\tau/2} - e^{j2\pi(f-mF)\tau/2}}{-j2\pi(f-mF)}\right] = \cdots \\ \cdots &= \lim_{\tau\to\infty}\tau\left[\frac{\sin((\pi(f-mF)\tau)}{\pi(f-mF)\tau}\right] = \delta(f-mF) \end{aligned} \tag{A21}$$

The Dirac delta function in frequency arises in the limit due to the properties of the $\sin(x)/x = \mathrm{sinc}(x/\pi)$ function: it is equal to 1 where $x = 0$ (as a result of another limit) and in (A21) the zero-crossings collapse toward $f = mF$ in frequency, and tend to 0 for all other f. Consequently, the result is that (A20) becomes

$$\int_{-\infty}^{\infty} e^{-j2\pi ft}\frac{1}{T}\sum_{m=-\infty}^{\infty} e^{j2\pi mFt}dt = \frac{1}{T}\sum_{m=-\infty}^{\infty}\delta(f-mF) \tag{A22}$$

and via (A19), reproduces (A15) as required:

$$\int_{-\infty}^{\infty} e^{-j2\pi ft}\sum_{n=-\infty}^{\infty}\delta(t-nT)dt = \frac{1}{T}\sum_{m=-\infty}^{\infty}\delta(f-mF) \tag{A23}$$

References

1. Sheleg, B. A Matrix-Fed Circular Array for Continuous scanning. *Proc. IEEE* **1968**, *56*, 2016–2027. [CrossRef]
2. Davies, D.E.N. Electronic steering of multiple nulls for circular arrays. *Electron. Lett.* **1977**, *13*, 669–670. [CrossRef]
3. Rahim, T.; Davies, D.E.N. Effect of directional elements on the directional response of circular antenna arrays. *IEE Proc.* **1982**, *129*, 18–22. [CrossRef]

4. Davis, J.G.; Gibson, A.A.P. Phase Mode Excitation in Beamforming Arrays. In Proceedings of the 3rd European Radar Conference, Manchester, UK, 13–15 September 2006.
5. Padgett, M.J. Orbital angular momentum 25 years on [Invited]. *Opt. Express* **2017**, *25*, 11265. [CrossRef] [PubMed]
6. Trichili, A.; Park, K.; Zghal, M.; Ooi, B.S.; Alouini, M. Communicating Using Spatial Mode Multiplexing: Potentials, Challenges and Perspectives. *arXiv*, 2018; arXiv:1808.02462v2.
7. Edfors, O.; Johansson, A.J. Is Orbital Angular Momentum (OAM) Based Radio Communication an Unexploited Area? *IEEE Trans. Antennas Propag.* **2012**, *60*, 1126–1131. [CrossRef]
8. Cagliero, A.; de Vita, A.; Gaffoglio, R.; Sacco, B. A New Approach to the Link Budget Concept for an OAM Communication Link. *IEEE Antennas Propag. Lett.* **2016**, *15*, 568–571. [CrossRef]
9. Zhang, C.; Ma, L. Millimetre wave with rotational orbital angular momentum. *Sci. Rep.* **2016**, *6*, 31921. [CrossRef] [PubMed]
10. Zhang, C.; Ma, L. Detecting the orbital angular momentum of electromagnetic waves using virtual rotational antenna. *Sci. Rep.* **2017**, *7*, 4585. [CrossRef] [PubMed]
11. Zhao, Z.; Yan, Y.; Xie, G.; Ren, Y.; Ahmed, N.; Wang, Z.; Liu, C.; Willner, A.J.; Song, P.; Hashemi, H.; et al. A Dual-Channel 60 GHz Communications Link Using Patch Antenna Arrays to Generate Data-Carrying Orbital-Angular-Momentum Beams. In Proceedings of the 2016 IEEE International Conference on Communications (ICC), Kuala Lumpur, Malaysia, 22–27 May 2016.
12. Tennant, A.; Allen, B. Generation of OAM radio waves using circular time-switched array antenna. *Electron. Lett.* **2012**, *48*, 1365–1366. [CrossRef]
13. Allen, B.; Tennant, A.; Bai, Q.; Chatziantoniou, E. Wireless data encoding and decoding using OAM modes. *Electron. Lett.* **2014**, *50*, 232–233. [CrossRef]
14. Cano, E.; Allen, B. Multiple-antenna phase-gradient detection for OAM radio communications. *Electron. Lett.* **2015**, *51*, 724–725. [CrossRef]
15. Chen, J.; Liang, X.; He, C.; Geng, J.; Jin, R. High-sensitivity OAM phase gradient detection based on time-modulated harmonic characteristic analysis. *Electron. Lett.* **2017**, *53*, 812–814. [CrossRef]
16. Drysdale, T.D.; Allen, B.; Stevens, C. Discretely-Sampled Partial Aperture Receiver for Orbital Angular Momentum Modes. In Proceedings of the 2017 IEEE International Symposium on Antennas and Propagation & USNC/URSI National Radio Science Meeting, San Diego, CA, USA, 9–14 July 2017; pp. 1431–1432.
17. Hu, Y.; Zheng, S.; Zhang, Z.; Chi, H.; Jin, X.; Zhang, X. Simulation of orbital angular momentum radio communication systems based on partial aperture sampling receiving scheme. *Inst. Eng. Technol. (IET) J. IET Microw. Antennas Propag.* **2016**, *10*, 1043–1047. [CrossRef]
18. Drysdale, T.D.; Allen, B.; Okon, E. Sinusoidal Time-Modulated Uniform Circular Array for Generating Orbital Angular Momentum Modes. In Proceedings of the IEEE 11th European Conference on Antennas and Propagation (EUCAP), Paris, France, 19–24 March 2017.
19. Laakso, T.I.; Välimäki, V.; Karjalainen, M.; Laine, U.K. Splitting the Unit Delay. *IEEE Signal Process. Mag.* **1996**, *13*, 30–60. [CrossRef]
20. Rahman, M.; Jahromi, M.N.; Mirjavadi, S.S.; Hamouda, A.M. Compact UWB Band-Notched Antenna with Integrated Bluetooth for Personal Wireless Communication and UWB Applications. *Appl. Sci. Electron.* **2019**, *8*, 158. [CrossRef]

Article

Mode-Selective Photonic Lanterns for Orbital Angular Momentum Mode Division Multiplexing

Yan Li [1,*], Yang Li [1], Lipeng Feng [1], Chen Yang [2], Wei Li [1], Jifang Qiu [1], Xiaobin Hong [1], Yong Zuo [1], Hongxiang Guo [1], Weijun Tong [2] and Jian Wu [1,*]

1 State Key Laboratory of Information Photonics and Optical Communications, Beijing University of Posts and Telecommunications, Beijing 100876, China; aslan_ly@163.com (Y.L.); 15201017279@163.com (L.F.); w_li@bupt.edu.cn (W.L.); jifangqiu@bupt.edu.cn (J.Q.); xbhong@bupt.edu.cn (X.H.); yong_zuo@bupt.edu.cn (Y.Z.); hxguo@bupt.edu.cn (H.G.)

2 State Key Laboratory of Optical Fibre and Cable Manufacture Technology, Yangtze Optical Fibre and Cable Joint Stock Limited Company, Wuhan 430074, China; chenyang@yofc.com (C.Y.); tongweijun@yofc.com (W.T.)

* Correspondence: liyan1980@bupt.edu.cn (Y.L.); jianwu@bupt.edu.cn (J.W.)

Received: 11 April 2019; Accepted: 27 May 2019; Published: 30 May 2019

Abstract: We analyze the mode evolution in mode-selective photonic lanterns with respect to taper lengths, affected by possible mode phase differences varying along the taper. As a result, we design a three-mode orbital angular momentum (OAM) mode-selective photonic lantern by optimizing the taper length with mode crosstalk below −24 dB, which employs only one single mode fiber port to selectively generate one OAM mode.

Keywords: mode division multiplexing; orbital angular momentum; photonic lantern

1. Introduction

Mode division multiplexing (MDM) has attracted great interest due to its potential for addressing the forthcoming capacity crunch [1]. The orbital angular momentum (OAM) modes can be extensively used as separate channels in MDM communication networks due to the orthogonality [2]. As the key techniques, the generation and multiplexing of OAM modes in MDM systems have been achieved in free space [3], silicon photonics [4], and fiber-based configurations [5]. Due to the high insertion loss, high fabrication cost, and limited scalability, the free space and silicon photonics-based solutions are still subject to further improvements and upgrading. The fiber-based solutions play a promising role to perform OAM manipulations because of their compatibility with transmission fibers and compactness. They can be widely applied in practical communication systems. Some fiber-based solutions for OAM mode generation have been carried out [5–7], but a simple and feasible fiber-based OAM multiplexing method is still yet to be developed.

Meanwhile, all-fiber-based techniques, e.g., photonic lanterns (PLs), which offer the potential for direct integration with existing telecom/datacom infrastructures and low insert loss, are highly desirable [8]. A PL can be nearly lossless, scaled to many modes, and robust because it can be directly spliced to one few-mode fiber (FMF) and several single-mode fibers (SMFs) with the right design [9]. Previous work [9] demonstrated a 3 × 1 fiber-based photonic lantern spatial multiplexer with mode selectivity greater than 6 dB and transmission loss of less than 0.3 dB. PLs for linear polarized (LP) mode multiplexing have been reported, which can support 10 LP modes so far [10].

Recently, OAM mode division multiplexing in the systems containing PLs has been reported [11,12]. In [11], an OAM mode multiplexer using an annular core mode-selective photonic lantern (MSPL) is proposed. However, it cannot simultaneously multiplex both OAM_{-1} and OAM_{+1}, as it has to employ two input SMF ports of a MSPL to generate one OAM mode. We have previously proposed a one-port excited all-fiber OAM multiplexer based on cascading a MSPL and a mode-polarization controller [12], but the mode-polarization controller introduces instability and additional complexity.

In this paper, we analyze the periodic evolution of the PL output modes with respect to taper lengths, affected by possible mode phase differences varying along the taper. Then, we design a three-mode orbital angular momentum mode-selective photonic lantern (OAM-MSPL) utilizing the phase difference caused by the mode degeneracy breaking along a specific area of the taper. The OAM_0, OAM_{-1}, and OAM_{+1} can be excited respectively when selectively and separately injecting light into the three SMF cores. Simulation results show that the mode crosstalk of the OAM-MSPL is below −24 dB.

2. Principle

A taper transition can couple light between one FMF and several SMFs. If the number of SMFs matches the number of spatial modes in the FMF, the transition can have low loss in both directions [13]. Taking a three-mode PL as an example, a standard PL is fabricated by adiabatically tapering three separate SMFs in a low-index glass jacket, as shown in Figure 1a. As unitary coupling between SMFs and one FMF is only possible by optimizing the arrangements of the cores [14], the cross section of the untapered end of the PL is shown in Figure 1b, which can be considered approximately as three isolated fiber cores (Fiber 1, 2, and 3 cores) with a circular cladding and a low-index jacket. The structure is tapered such that the SMF cores nearly vanish, the SMF cladding becomes the new few-mode core, and the low-index capillary becomes the FMF cladding. The tapered end of the PL matches the common two-mode fiber supporting LP_{01} and LP_{11} modes. The original LP_{01} modes in the SMF cores eventually evolve into the LP_{01} and LP_{11} modes in the FMF core.

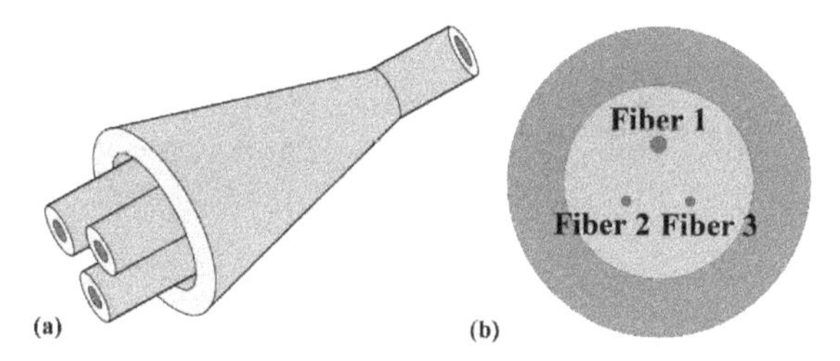

Figure 1. The (**a**) structure and (**b**) cross section of a photonic lantern (PL). Note: (**b**) is not to scale.

In this paper, we obtain the eigenmode profiles and effective refractive indexes of the PL at different positions along the taper, simulated by mode analysis using the finite element method (FEM). However, FEM is not able to simulate the many elements of the light transmission in the PL, such as the adiabatic taper criterion and the actual forms of the excited supermodes, i.e., local modes. It is not able to analyze the possible effects brought by the variation of the taper lengths either. Therefore, as a supplement, we use the beam propagation method (BPM) to analyze the evolution of the local modes excited by selectively injecting light into the SMF cores during the taper of the PL, in order to consider the variation of the light transmission conditions.

Three-mode MSPLs for LP mode multiplexing can be divided into two types: the standard three-mode MSPL designed with three totally different SMF cores, and the three-mode mode-group-selective PL owning a pair of symmetric SMF cores with a lower initially designed normalized frequency value (V value) compared to the other core. We simulate both these types to analyze the evolution of the eigenmodes and the local modes using FEM and BPM. In this paper, the simulated PL has three cores arranged in an equilateral triangle shape with a core pitch of 42.0 μm, as the reduced cladding fibers can be used to assist adiabaticity [15,16]. The core index is 1.4482. For the standard MSPL, the core diameters are 11.0 μm, 8.65 μm, and 6.55 μm, respectively. For the mode-group-selective PL, the core diameters are 11.0 μm, 6.55 μm, and 6.55 μm, as the Fiber 2 core and the Fiber 3 core are identical. The cladding and jacket diameters are 125 and 1116 μm, and the cladding and jacket indexes are 1.444 and 1.4398, respectively.

As shown in Figure 2a, for the standard MSPL, the original eigenmodes of the untapered end are the LP_{01} modes in the three separate SMF cores. The second- and third-highest indexes remain separate until the SMF cores nearly vanish and the local modes evolve into the degenerate LP_{11} modes. When selectively injecting light into the Fiber 2 or 3 core, as shown in Figure 2b, at the same taper ratio, the local mode shown as the BPM profile is in correspondence with the eigenmode, shown as the FEM profile. On account of the separation of the effective refractive indexes, the second- and third-order local modes individually evolves into the two second-order eigenmodes in the FMF end, i.e., LP_{11a} and LP_{11b}.

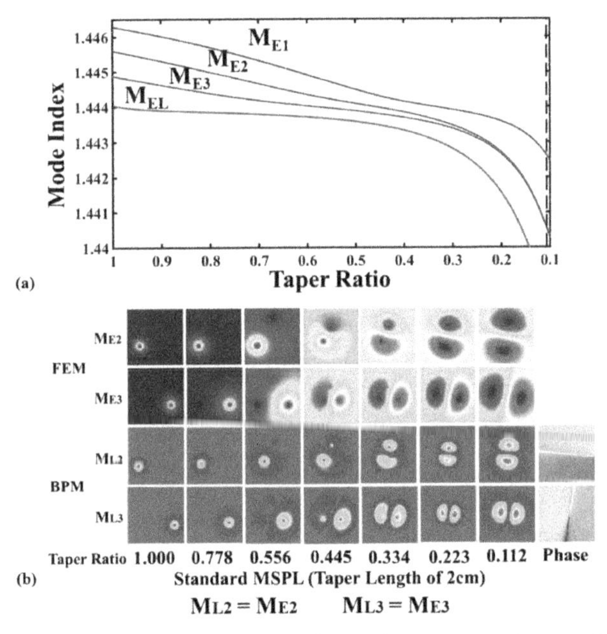

Figure 2. (**a**) The FEM simulation mode indexes evolution along the taper of the standard MSPL. M_{E1}, M_{E2}, and M_{E3} represent the first- and the two second-order eigenmodes. The M_{EL} represents the leaky mode simulated by FEM. The black dashed line indicates the taper ratio of 0.112. (**b**) The comparison of the FEM/BPM simulated mode profiles at different positions along the taper when selectively injecting light into the Fiber 2 and 3 cores of the standard MSPL. M_{L1}, M_{L2}, and M_{L3} represent the local modes when individually injecting light into Fiber 1, 2, or 3 cores, respectively. BPM: beam propagation method; FEM: finite element method; MSPL: mode-selective photonic lantern.

As shown in Figure 3a, for the mode-group-selective PL, as the Fiber 2 and 3 cores share the same V value, the initial eigenmodes are the in-phase and reverse-phase combination of the LP_{01} modes in the Fiber 2 and 3 cores, and their effective refractive indexes are identical at the starting area of the taper. When selectively injecting light into either of the pair of the totally symmetric and identical cores (Fiber 2 or 3 cores), both the degenerate second-order eigenmodes are excited. The two eigenmodes break the degeneracy and bring the difference value of the effective refractive indices in the deciding area of the taper (where the effective index differences of M_{L2} and M_{L3} are larger than 10^{-4}, the value normally considered as the floor in order to eliminate MIMO-DSP in PLs [17]), then they return to degeneracy again at the end of the taper. As shown in Figure 3b, each of the local modes shown as the BPM profile is the isometric superposition of the two eigenmodes shown as FEM profiles, and the composition of the output mode in the FMF end depends on the accumulation of the phase differences brought by the two eigenmodes in the deciding area of the taper. The relationship can be described as follows:

$$M_{L2} = M_{E2} + \exp(i\varphi)M_{E3} \quad (1)$$

$$M_{L3} = M_{E2} + \exp[i(\pi-\varphi)]M_{E3}, \quad (2)$$

where M_{L2} and M_{L3} represent the local modes when individually injecting light into the Fiber 2 or 3 core, M_{E2} and M_{E3} represent the two second-order eigenmodes, and φ represents the final phase difference between the two parts of the excited eigenmodes generated during the taper. It is already known that one pair of degenerated LP_{11} modes which satisfy the $\pi/2$ phase difference can be used to generate $OAM_{\pm 1}$ modes in optical fibers [7], as follows:

$$OAM_{-1} = LP_{11a} + i\, LP_{11b} \quad (3)$$

$$OAM_{+1} = LP_{11a} - i\, LP_{11b}. \quad (4)$$

Therefore, when selectively injecting light into the Fiber 2 or 3 core, if φ equals $\pi/2$ (or $-\pi/2$), the output mode is one of the $OAM_{\pm 1}$ modes, and if φ equals 0 (or π), the output mode is one of the LP_{11} modes.

The BPM simulation mode profiles of the PLs with lengths of 2 cm, 2.65 cm, and 3.3 cm are shown in Figure 3b–d, as the final taper ratio is determined as 0.112. As described above, when injecting light into either the Fiber 2 or 3 core selectively, the variation of the phase differences leads to the changing of the superposed output mode. For instance, in the simulated PL with a taper length of 2 cm, the excited mode of the Fiber 2 core corresponds to OAM_{-1}. It becomes LP_{11a} in the simulated PL with a taper length of 2.65 cm, and then turns to OAM_{+1} when the taper length is 3.3 cm. The output mode profiles of the PLs with different taper lengths are shown in Figure 4, based on the BPM simulation when injecting light into each of the three SMF cores.

It has been shown that there is a periodic variation of the generating $OAM_{\pm 1}$ modes at a half cycle of 1.3 cm as the taper length increases. Hence, we can obtain a designed OAM-MSPL if the LP_{11} modes satisfies the $\pi/2$ phase difference. Moreover, we could also optimize taper ratios and V values of the SMF cores to control the phase difference variation in order to obtain low-loss OAM-MSPLs.

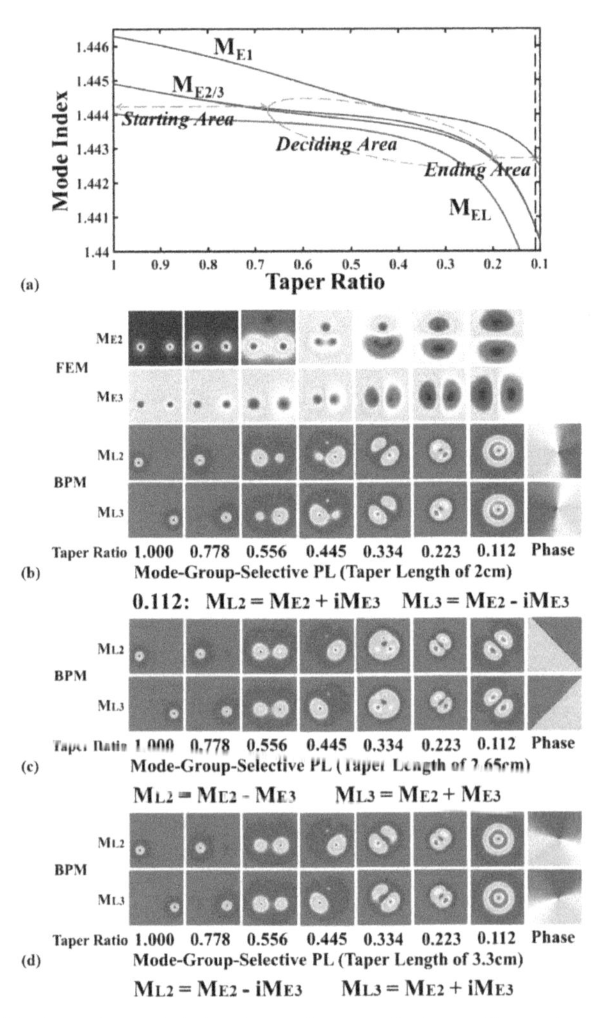

Figure 3. The FEM simulation mode indexes' evolution along the taper of the mode-group-selective PL (**a**). The comparison of the FEM/BPM simulated mode profiles at different positions along the taper when selectively injecting light into the Fiber 2 and 3 cores of the mode-group-selective PL with the taper length of (**b**) 2, (**c**) 2.65, and (**d**) 3.3 cm.

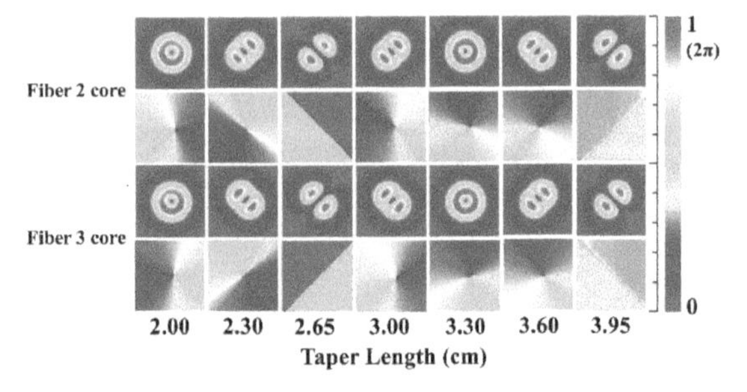

Figure 4. The BPM simulation output mode profiles of the PLs with different taper lengths when selectively injecting light into the Fiber 2 and 3 cores.

3. Simulation Results

As a result of the mode evolution analysis in MSPLs described above, we design the OAM-MSPL with a taper length of 2.0 cm, similar to practical PLs. When injecting light into the Fiber 1 core, the mode coupling efficiencies of OAM_0, OAM_{-1}, and OAM_{+1} are 99.372%, 0.306%, and 0.306%, respectively, which were obtained by calculating the correlation coefficients between the target modes of the PL and the matching FMF as the waist [18]. As for Fiber 2, they are 0.318%, 99.665%, and 0.001%, respectively. As for Fiber 3, they are 0.318%, 0.001%, and 99.665%, respectively. Simulation results show good characteristics of multiplexing when the few-mode ends of one pair of the designed PLs are connected, as shown in Figure 5a.

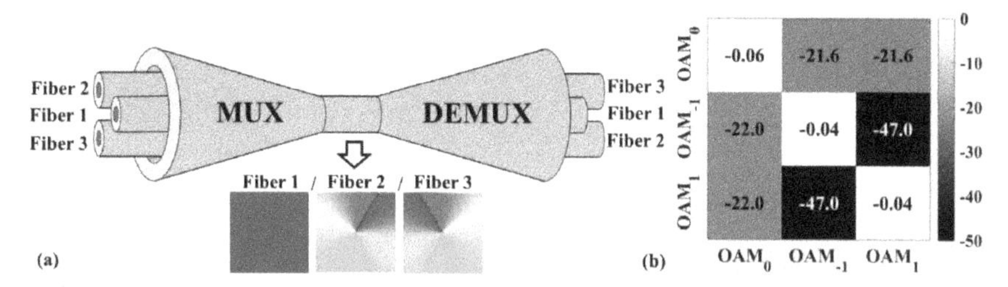

Figure 5. (**a**) The schematic of the multiplexer/de-multiplexer (MUX/DEMUX) simulation. (**b**) Calculated mode crosstalk matrix of one pair of the OAM-MSPLs during selective excitation (bottom axis) and measurement (left axis). The diagonal matrix elements represent the simulated insertion loss of one pair of the OAM-MSPLs. OAM-MSPLs: orbital angular momentum mode-selective photonic lanterns.

When injecting light into the target fiber core of the MUX-PL (where light is injected into the SMF ends and is emitted from the FMF end), we obtain the mode crosstalk, which is defined as the ratio of power in the corresponding fiber core of the DEMUX-PL (where light is injected into the FMF end and is emitted from the SMF ends) over power in the other cores. The matrix elements shown in Figure 5b are displayed in units of decibels (dB), with the mode crosstalk of one pair of the designed PLs below −21 dB, based on the symmetric MUX/DEMUX simulation structure. The individual simulation shows the mode crosstalk of the single PL is below −24 dB, which matches the result of the MUX/DEMUX simulation. The diagonal elements represent the simulated insertion loss (IL) of one pair of the designed PLs, which is below −0.06 dB, as the IL of a single PL is below −0.03 dB on average.

We calculate correlation coefficients between the target OAM modes and FMF via the overlap integral method by scanning the linear taper lengths and the light wavelengths, as shown in Figure 6a,d. With the mode crosstalk shown in Figure 6b,c, it is indicated that the OAM-MSPL has a taper length error tolerance longer than 1 mm, which can be controlled by common PL-tapering processors. The mode crosstalk of a single PL can be controlled below −21 dB, with the IL below −0.06 dB. Furthermore, as simulations performed, the designed PL works stably in the whole C-band and L-band with low loss and crosstalk.

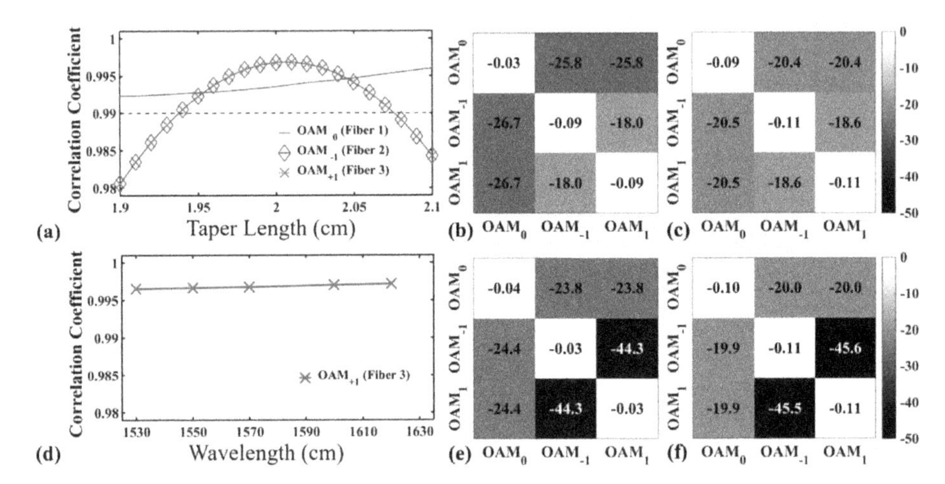

Figure 6. Calculated correlation coefficients of OAM-MSPL with (**a**) taper lengths around 2 cm when injecting light into all of the cores and (**d**) with wavelengths in the C-band and L-band when injecting light into the Fiber 3 core. Calculated mode crosstalk matrices of pairs of the PLs with taper lengths of (**b**) 1.95 cm and (**c**) 2.05 cm, or with wavelengths at (**e**) 1530 nm and (**f**) 1620 nm.

4. Conclusions

In conclusion, we analyze the periodic evolution of the PL output modes with respect to taper lengths, affected by possible mode phase differences varying along the taper. Then, we design a three-mode OAM-MSPL for MDM with mode crosstalk below −24 dB and IL below −0.03 dB. As the simulations performed revealed, the designed OAM-MSPL has a taper length error tolerance longer than 1 mm with mode crosstalk below −21 dB and IL below −0.06 dB, and works stably in the whole C-band and L-band with low loss and crosstalk.

Author Contributions: This paper was built on original ideas by Y.L. (Yan Li) and Y.L. (Yang Li), and both of them performed the calculations. All authors contributed to the writing and editing of the manuscript.

Funding: This research was funded by the National Natural Science Foundation of China (NSFC) (61875019, 61675034, 61875020, 61571067), The Fund of State Key Laboratory of IPOC (BUPT), and The Fundamental Research Funds for the Central Universities.

Conflicts of Interest: The authors declare no conflict of interest.

References

1. Essiambre, R.J.; Kramer, G.; Winzer, P.J.; Foschini, G.J.; Goebel, B. Capacity Limits of Optical Fiber Networks. *J. Lightwave Technol.* **2010**, *28*, 662–701. [CrossRef]
2. Bozinovic, N.; Yue, Y.; Ren, Y.; Tur, M.; Kristensen, P.; Huang, H.; Willner, A.E.; Ramachandran, S. Terabit-scale orbital angular momentum mode division multiplexing in fibers. *Science* **2013**, *340*, 1545–1548. [CrossRef] [PubMed]
3. Gregg, P.; Mirhosseini, M.; Rubano, A.; Marrucci, L.; Karimi, E.; Boyd, R.; Ramachandran, S. Q-plates for Switchable Excitation of Fiber OAM Modes. In Proceedings of the 2015 Conference on Lasers and Electro-Optics (CLEO), San Jose, CA, USA, 10–15 May 2015; pp. 1–2.
4. Cai, X.; Wang, J.; Strain, M.J.; Johnson-Morris, B.; Zhu, J.; Sorel, M.; O'Brien, J.L.; Thompson, M.G.; Yu, S. Integrated compact optical vortex beam emitters. *Science* **2012**, *338*, 363–366. [CrossRef] [PubMed]
5. Pidishety, S.; Pachava, S.; Gregg, P.; Ramachandran, S.; Brambilla, G.; Srinivasan, B. Orbital angular momentum beam excitation using an all-fiber weakly fused mode selective coupler. *Opt. Lett.* **2017**, *42*, 4347–4350. [CrossRef] [PubMed]

6. Zeng, X.; Li, Y.; Mo, Q.; Tian, Y.; Li, W.; Liu, Z.; Wu, J. Experimental demonstration of Compact and Robust All-Fiber Orbital Angular Momentum Generator. In Proceedings of the 2016 European Conference on Optical Communication (ECOC), Dusseldorf, Germany, 18–22 September 2016; pp. 1–3.
7. Li, S.; Mo, Q.; Hu, X.; Du, C.; Wang, J. Controllable all-fiber orbital angular momentum mode converter. *Opt. Lett.* **2015**, *40*, 4376–4379. [CrossRef] [PubMed]
8. Pidishety, S.; Khudus, M.A.; Gregg, P.; Ramachandran, S.; Srinivasan, B.; Brambilla, G. OAM Beam Generation using All-fiber Fused Couplers. In Proceedings of the 2016 Conference on Lasers and Electro-Optics (CLEO), San Jose, CA, USA, 5–10 June 2016; pp. 1–2.
9. Leon-Saval, S.G.; Fontaine, N.K.; Salazar-Gil, J.R.; Ercan, B.; Ryf, R.; Bland-Hawthorn, J. Mode-selective photonic lanterns for space-division multiplexing. *Opt. Express* **2014**, *22*, 1036–1044. [CrossRef] [PubMed]
10. Alvarado-Zacarias, J.C.; Huang, B.; Eznaveh, Z.S.; Fontaine, N.K.; Chen, H.; Ryf, R.; Antonio-Lopez, J.E.; Correa, R.A. Experimental analysis of the modal evolution in photonic lanterns. In Proceedings of the 2017 Optical Fiber Communications Conference and Exhibition (OFC), Los Angeles, CA, USA, 19–23 March 2017; pp. 1–3.
11. Eznaveh, Z.S.; Zacarias, J.C.; Lopez, J.E.; Shi, K.; Milione, G.; Jung, Y.; Thomsen, B.C.; Richardson, D.J.; Fontaine, N.; Leon-Saval, S.G.; et al. Photonic lantern broadband orbital angular momentum mode multiplexer. *Opt. Express* **2018**, *26*, 30042–30051. [CrossRef] [PubMed]
12. Zeng, X.; Li, Y.; Feng, L.; Wu, S.; Yang, C.; Li, W.; Tong, W.; Wu, J. All-Fiber OAM Mode Multiplexer Employing Photonic Lantern and Mode-Polarization Controller. In Proceedings of the 2018 European Conference on Optical Communication (ECOC), Rome, Italy, 23–27 September 2018; pp. 1–3.
13. Leon-Saval, S.G.; Birks, T.A.; Bland-Hawthorn, J.; Englund, M. Multimode fiber devices with single-mode performance. *Opt. Lett.* **2005**, *30*, 2545–2547. [CrossRef] [PubMed]
14. Fontaine, N.K.; Ryf, R.; Bland-Hawthorn, J.; Leon-Saval, S.G. Geometric requirements for photonic lanterns in space division multiplexing. *Opt. Express* **2012**, *20*, 27123–27132. [CrossRef] [PubMed]
15. Yeirolatsitis, S.; Harrington, K.; Thomson, R.R.; Birks, T.A. Mode-selective photonic lanterns from multicore fibres. In Proceedings of the 2017 Optical Fiber Communications Conference and Exhibition (OFC), Los Angeles, CA, USA, 19–23 March 2017; pp. 1–3.
16. Shen, L.; Gan, L.; Huo, L.; Yang, C.; Tong, W.; Fu, S.; Tang, M.; Liu, D. Design of highly mode group selective photonic lanterns with geometric optimization. *Appl. Opt.* **2018**, *57*, 7065–7069. [CrossRef] [PubMed]
17. Ramachandran, S.; Fini, J.M.; Mermelstein, M.; Nicholson, J.W.; Ghalmi, S.; Yan, M.F. Ultra-large effective-area, higher-order mode fibers: A new strategy for high-power lasers. *Laser Photonics Rev.* **2008**, *2*, 429–448. [CrossRef]
18. Sai, X.; Li, Y.; Yang, C.; Li, W.; Qiu, J.; Hong, X.; Zuo, Y.; Guo, H.; Tong, W.; Wu, J. Design of elliptical-core mode-selective photonic lanterns with six modes for MIMO-free mode division multiplexing systems. *Opt. Lett.* **2017**, *42*, 4355–4358. [PubMed]

Article

Holographic Silicon Metasurfaces for Total Angular Momentum Demultiplexing Applications in Telecom

Gianluca Ruffato [1,2], Michele Massari [2,3], Pietro Capaldo [3] and Filippo Romanato [1,2,3,*]

1. Department of Physics and Astronomy 'G. Galilei', University of Padova, via Marzolo 8, 35131 Padova, Italy; gianluca.ruffato@unipd.it
2. Laboratory for Nanofabrication of Nanodevices (LaNN), EcamRicert, Corso Stati Uniti 4, 35127 Padova, Italy; massari@iom.cnr.it
3. CNR-INFM TASC IOM National Laboratory, S.S. 14 Km 163.5, 34012 Trieste, Italy; capaldo@iom.cnr.it

* Correspondence: filippo.romanato@unipd.it

Received: 20 May 2019; Accepted: 6 June 2019; Published: 11 June 2019

Featured Application: Generation and sorting of optical beams carrying orbital angular momentum of light for combined polarization- and mode-division multiplexing in the telecom infrared, either for free-space or multi-mode fiber transmission.

Abstract: The simultaneous processing of orbital angular momentum (OAM) and polarization has recently acquired particular importance and interest in a wide range of fields ranging from telecommunications to high-dimensional quantum cryptography. Due to their inherently polarization-sensitive optical behavior, Pancharatnam–Berry optical elements (PBOEs), acting on the geometric phase, have proven to be useful for the manipulation of complex light beams with orthogonal polarization states using a single optical element. In this work, different PBOEs have been computed, realized, and optically analyzed for the sorting of beams with orthogonal OAM and polarization states at the telecom wavelength of 1310 nm. The geometric-phase control is obtained by inducing a spatially-dependent form birefringence on a silicon substrate, patterned with properly-oriented subwavelength gratings. The digital grating structure is generated with high-resolution electron beam lithography on a resist mask and transferred to the silicon substrate using inductively coupled plasma-reactive ion etching. The optical characterization of the fabricated samples confirms the expected capability to detect circularly-polarized optical vortices with different handedness and orbital angular momentum.

Keywords: Pancharatnam–Berry optical elements; silicon metasurfaces; mode division multiplexing; orbital angular momentum; polarization division multiplexing; electron beam lithography; subwavelength digital gratings; nanofabrication; reactive ion etching

1. Introduction

In the last decades, the possibility to structure the spatial degree of freedom of light has acquired increasing interest, with applications in a wide range of fields. In particular, the exploitation of light beams with helical phase-fronts has provided disruptive achievements in microscopy [1,2], astronomy [3], particle manipulation [4], holography [5], and information and communication technology (ICT) [6,7]. Since the seminal paper of Allen and coworkers [8] demonstrated that such beams carry orbital angular momentum (OAM), the study on methods and devices to generate and control this still unexploited degree of freedom has given rise to a flourishing research field [9]. It is especially in the ICT that the orbital angular momentum of light has demonstrated the most promising applications, in combination with other degrees of freedom of light [10]. As a matter of fact, the OAM degree of freedom opens to an unbounded state space, in which light beams carrying different integer

OAM values are orthogonal to each other and can be exploited for the transmission of different data streams at the same frequency with no interference [11]. The aggregate combination of OAM-mode division multiplexing (OAM-MDM) with other well-established multiplexing methods, e.g., time, polarization, wavelength, and amplitude/phase, has demonstrated to provide a significant increase in the spectral efficiency of today's optical networks [12], both in free-space [13] and optical fibers [14,15], offering a solution to the problem of optical network saturation [16]. Both in the classical and quantum regimes, the combined manipulation of OAM and polarization has acquired paramount importance, and novel devices are required for the parallel detection in a compact and effective way. As a matter of fact, optical vortices propagating in multi-mode fibers have been demonstrated to be circularly polarized [15]. In the single-photon regime, the combination of orbital angular momentum and polarization opens to a wider state space for quantum-key distribution (QKD) applications [17], in which higher security and robustness against errors and eavesdropping are guaranteed with respect to standard protocols limited to polarization [7]. Novel formulations and innovative implementations of standard QKD protocols have been developed and demonstrated [18], both in discrete variable (DV-QKD) and continuous variable (CV-QKD) scenarios [19]. Concurrently, research efforts have focused on the design and realization of polarization-sensitive OAM (de)multiplexers, in order to generate and sort the state space exploited for high-dimensional QKD [20,21].

In the last decade, several methods have been conceived for demultiplexing, i.e., the separation, of a superposition of beams with different values of OAM [22]. In particular, increasing interest has been devoted to solutions which could offer high miniaturization and integration levels, fabrication protocols suitable for mass-production, and backward compatibility with different multiplexing techniques. In order to improve the miniaturization level, we recently disclosed the realization of 3D multi-level phase-only diffractive optical elements [23–26] performing OAM-mode generation and detection in the visible range, based either on *log-pol* optical transformation [27] or OAM-mode projection [28,29]. In comparison with bulky refractive elements, the diffractive version provides a miniaturized and almost flat implementation, in particular, when shorter focal lengths are necessary, i.e., for high miniaturization. On the other hand, the design of diffractive optics turns out to be optimized within a narrow bandwidth, therefore they exhibit a decrease in efficiency when operating far from the optimal wavelength. The optical thickness is inversely proportional to the refractive index of the material, and increases proportionally to the design wavelength. If the transparency of silicon in the telecom infrared suggests the exploitation of this high-refractive index material in order to further reduce the optical thickness, then, on the other hand, the fabrication of 3D surface-relief patterns in silicon is still undoubtedly challenging.

An alternative method for phase-fronts manipulation is provided by Pancharatnam–Berry optical elements (PBOE) acting on the geometric phase of light. Unlike refractive and diffractive optics, in PBOEs the phase change is not produced by means of an optical path difference, but is the result of a space-variant modification of the polarization state of light [30]. This is achieved by realizing an artificial material, i.e., a metasurface, which is both inhomogeneous and anisotropic, in order to create an effective anisotropic medium whose extraordinary axis orientation is spatially variant. The phase transferred to the input beam is equal to twice the value of the fast-axis orientation, therefore by properly engineering the anisotropy pattern it is possible to reshape the input phase-front in the desired way. With respect to conventional optics, the approach with metasurfaces can offer greater advantages owing to their digital profile and fixed thickness. In comparison with diffractive optics, metasurfaces show a broader band [31], since the wave-front is tailored by the geometric pattern of the optical element. In addition, since the optical response becomes inherently dependent on the input polarization [32], polarization-division multiplexing (PDM) can be easily implemented without the need of additional optics [33].

In this paper, we present the design and realization of sorting optics for OAM-MDM, in the form of Pancharatnam–Berry optical elements in silicon for the telecom wavelength of 1310 nm. We considered the demultiplexing method based on OAM-mode projection and we computed and realized different metasurfaces in silicon, performing both OAM-MDM and PDM. Despite its lower efficiency with respect to other methods, this technique allows to customize the channel constellation and the sorting OAM range, depending on the desired application. Different OAM sets and far-field channel configurations have been selected and presented, in order to demonstrate the versatility offered by this demultiplexing method.

The birefringence of the single PBOE subunit has been achieved artificially by structuring the silicon substrate with a digital subwavelength grating, which is experienced by the impinging wave as a uniaxial crystal whose fast axis is perpendicular to the grating ridges [34]. The resist mask fabricated on the silicon surface with high-resolution electron-beam lithography (EBL) was transferred to the substrate by a finely-tuned inductively coupled plasma—reactive plasma etching (ICP-RIE) process. The optical tests at the wavelength of 1310 nm, in the telecom O-band, confirm the expected capability of the designed optics to detect correctly input beams with different circular-polarization states and orbital angular momentum values.

2. Materials and Methods

2.1. Phase Pattern Calculation

A diffractive optical element designed to analyze the OAM spectrum in the set of OAM values $\{\ell_j\}$ presents a phase pattern $\Omega(u,v)$ which is given by the linear superposition of n orthogonal OAM modes $\{\psi_j\}$ as follows [35]:

$$\Omega(u,v) = \arg\left\{\sum_{j=1}^{n} c_j \psi_j^* \exp\left[i\alpha_j u + i\beta_j v\right]\right\}, \tag{1}$$

being $\{\psi_j = R_j(\rho,\vartheta)\exp(i\ell_j\vartheta)\}$, where $\vartheta = \arctan(v/u)$, $\rho = \sqrt{u^2+v^2}$, $\{R_j(\rho,\vartheta)\}$ describe the field spatial distributions and depend on the family of modes. $\{(\alpha_j, \beta_j)\}$ are the n vectors of carriers spatial frequencies in Cartesian coordinates, and $\{c_j\}$ are complex coefficients whose modulus is usually unitary, and the phases are fitted so that Equation (1) is an exact equality [23]. The set of parameters $\{c_j\}$ is calculated with the following integral:

$$c_j = \int_{-\infty}^{+\infty} du \int_{-\infty}^{+\infty} \psi_j \exp(i\Omega) \exp\left(-i\alpha_j u - i\beta_j v\right) dv, \tag{2}$$

The diffractive element is basically a computer-generated hologram originated from the linear combination of n fork-holograms. Each term in Equation (1) is given by the interference pattern of the jth OAM-mode with azimuthal phase term $\exp(i\ell_j\vartheta)$ with the tilted plane-wave $\exp(i\alpha_j u + i\beta_j v)$ defined by the corresponding carrier frequency. In the Fourier plane, the carrier frequencies manifest as separate spatial coordinates $\{(x_j, y_j)\}$ given by:

$$\begin{array}{l} x_j = \alpha_j \frac{f}{k} \\ y_j = \beta_j \frac{f}{k} \end{array}, \tag{3}$$

being f the focal length of the lens which is used for far-field reconstruction in f-f configuration, and $k = 2\pi/\lambda$, where λ is the working wavelength. When the optical element is illuminated with an integer OAM beam, the projection of the beam is optically performed over the selected OAM set, and a bright spot appears at the position corresponding to the input OAM value in far field (Figure 1) [23].

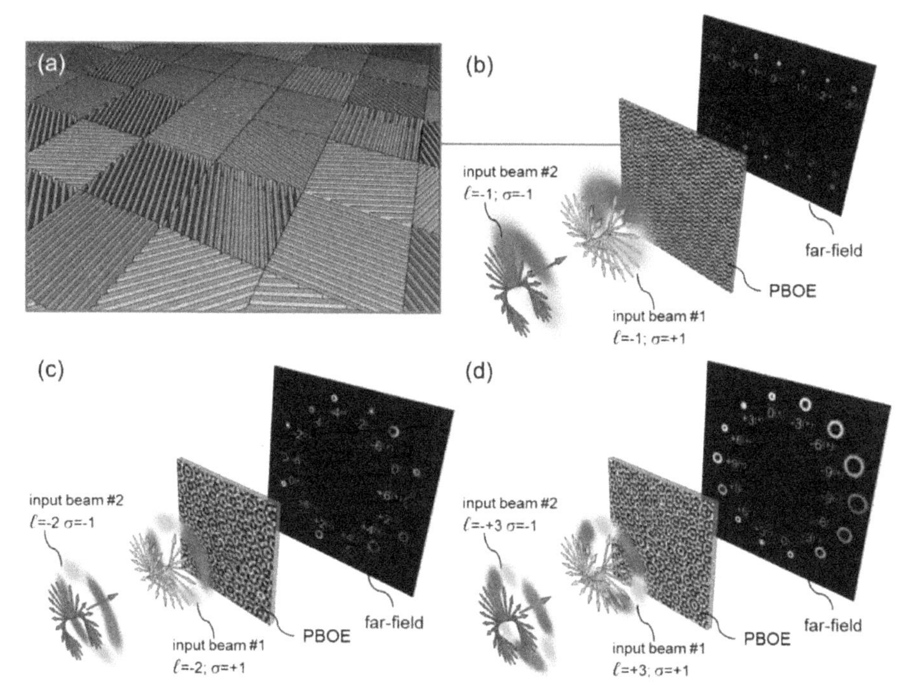

Figure 1. Scheme of the working principle of Pancharatnam–Berry optical elements (PBOEs)for orbital angular momentum (OAM) demultiplexing with the method of OAM-beam projection. If a circularly-polarized OAM-beam illuminates the optical element, a bright spot appears in the far field, at a position which depends on the polarization handedness and on the carried OAM. In the presented work, the PBOE has been fabricated in the form of a pixelated metasurface of rotated subwavelength gratings (**a**), and designed for the demultiplexing of 7 OAM channels and circular polarization states (14 channels in total) over different OAM sets, centered in $\ell = 0$, with increasing OAM separation $\Delta\ell = 1$ (**b**), $\Delta\ell = 2$ (**c**), $\Delta\ell = 3$ (**d**). Three different channel configurations have been considered and tested: Linear array (**b**), regular polygon (heptagon) (**c**), semicircle (**d**).

In a metasurface realization of the optical element, the phase pattern $\Omega(u,v)$ is obtained by fabricating an inhomogeneous and anisotropic effective medium, whose extraordinary-axis orientation $\theta(u,v)$ changes point-by-point and is equal to half the local phase value $\Omega(u,v)$. The two orthogonal circular polarizations exhibit a different behavior, as it follows [32]:

$$T(u,v)\begin{pmatrix} 1 \\ \pm i \end{pmatrix} = \cos\left(\frac{\delta}{2}\right)\begin{pmatrix} 1 \\ \pm i \end{pmatrix} - i\sin\left(\frac{\delta}{2}\right)\exp[\pm i\Omega(u,v)]\begin{pmatrix} 1 \\ \mp i \end{pmatrix} \tag{4}$$

being T the transmission matrix of the optical element, δ the phase delay between the ordinary and extraordinary axes of the metasurface effective medium, [1, $+i$] and [1, $-i$] the vectors of right-handed and left-handed circular polarizations in Jones matrix formalism, respectively (the normalization factor $1/\sqrt{2}$ has been omitted). In particular, when the metasurface is engineered in order to achieve the condition $\delta = \pi$ (π-delay between the two optical axes), the zero-order term is cancelled out and a total polarization conversion is obtained:

$$T\begin{pmatrix} 1 \\ \pm i \end{pmatrix} = -i\exp(\pm i\Omega)\begin{pmatrix} 1 \\ \mp i \end{pmatrix} \tag{5}$$

In this case, the two orthogonal circular polarizations experience opposite phase patterns:

$$\begin{aligned} \Omega^{(+)}(u,v) &= \arg\left\{ \sum_{j=1}^{n} c_j R_j^* \exp\left[-i\ell_j \vartheta + i\alpha_j u + i\beta_j v \right] \right\} \\ \Omega^{(-)}(u,v) &= \arg\left\{ \sum_{j=1}^{n} c_j R_j \exp\left[+i\ell_j \vartheta - i\alpha_j u - i\beta_j v \right] \right\} \end{aligned}, \quad (6)$$

and their corresponding sets of intensity spots appear at symmetric coordinates in far field:

$$\begin{aligned} x^{(-)}\left(\ell_j\right) &= -x^{(+)}\left(-\ell_j\right) \\ y^{(-)}\left(\ell_j\right) &= -y^{(+)}\left(-\ell_j\right) \end{aligned} \quad (7)$$

where the subscripts (+) and (−) stand for right-handed and left-handed circular polarizations, respectively.

As expressed by Equation (7), a beam carrying OAM equal to ℓ and right-handed circular polarization generates a bright spot at a position which is center-symmetric to the spot formed by the left-handed circularly-polarized state with opposite value of OAM. Hence, during the design of a metasurface performing demultiplexing over a properly-designed set of modes, particular attention should be paid to carefully choosing the spatial frequency carriers in order to prevent different channels from overlapping.

A custom code has been developed in MATLAB® in order to compute the phase patterns for the selected set $\{\ell_j\}$ of OAM values and the corresponding carriers frequencies $\{(\alpha_j, \beta_j)\}$. The implemented algorithm is based on a successive computation of the integrals in Equation (2) and of the sum in Equation (1), implementing the fast Fourier transform algorithm and applying precise constrains, as explained in [5,23], in particular phase quantization into 16 equally-spaced values in the range $[0, 2\pi)$.

In the following, three different configurations are presented and described. Each phase pattern performs the demultiplexing of circularly-polarized beams over 7 OAM values, for a total of 14 channels, with different OAM separation and far-field channel constellation: Linear array, regular polygon, semicircle.

2.1.1. Linear Array

We limited the choice to OAM values in the set from $\ell = -3$ to $\ell = +3$ for a total of 7 OAM values ($n = 7$). The spatial frequencies have been fixed in such a way that the far-field peaks were arranged along a line at equally spaced x-positions (see Figure 2b):

$$\begin{aligned} x_j^{(+)} &= \alpha\left(\tfrac{n+1}{2} - j\right)\tfrac{f}{k} = \alpha(4-j)\tfrac{f}{k} \\ y_j^{(+)} &= \beta\tfrac{f}{k} \end{aligned}, \quad (8)$$

where $j = 1, \ldots 7$. Considering Equation (7) and the symmetry of the far-field channels constellation, it results that:

$$\begin{aligned} x_j^{(-)} &= -\alpha(-4+j)\tfrac{f}{k} = x_j^{(+)} \\ y_j^{(-)} &= -\beta\tfrac{f}{k} = -y_j^{(+)} \end{aligned}, \quad (9)$$

The two orthogonal polarizations are therefore sorted over two distinct linear arrays without overlapping, as depicted in Figure 2b.

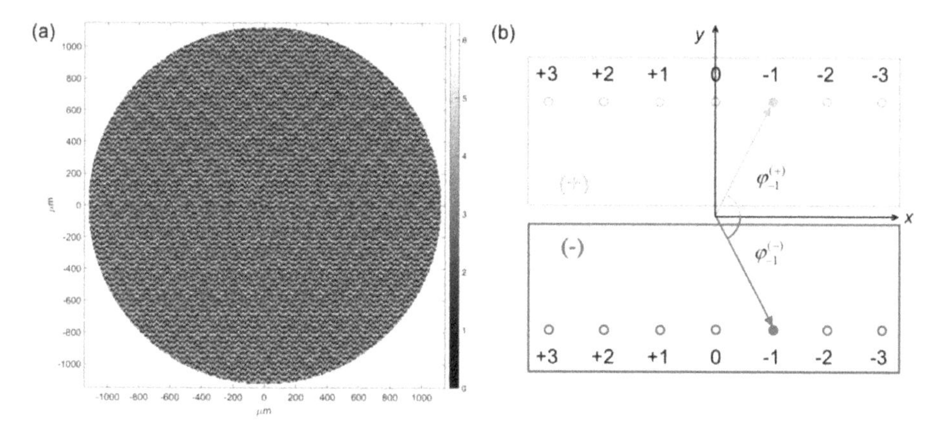

Figure 2. (**a**) Numerical phase pattern for the demultiplexing of optical beams with OAM in the set {−3, −2, −1, 0, +1, +2, +3} on a linear array. Pixel size: 6.250 µm × 6.250 µm. 16 phase levels. Radius length: 180 pixels. (**b**) Far-field channel scheme for the given OAM set and circular polarization states. Right-handed (in blue) and left-handed (in red) circularly-polarized OAM beams are detected in far field on two distinct linear arrays.

2.1.2. Regular Polygonal Configuration

We considered the OAM values in the set {−6, −4, −2, 0, +2, +4, +6} for a total of 7 OAM channels. We fixed the carrier spatial frequencies in order to arrange the far-field peaks at the vertices of a regular polygon, in the specific case a heptagon. In polar coordinates, the spatial frequencies are given by $\{(\rho_j, \theta_j)\} = \{(\gamma, j2\pi/7)\}$, $j = 1, \ldots 7$. Therefore, the far-field points appear at equally-spaced angular positions, specified as follows:

$$\begin{array}{l} r_j^{(+)} = r = \gamma \frac{f}{k} \\ \varphi_j^{(+)} = j\frac{2\pi}{n} = j\frac{2\pi}{7} \end{array}, \tag{10}$$

where $j = 1, \ldots 7$, being r the radius of the circumscribed circle. According to Equation (7), we have:

$$\tan\left[\varphi^{(-)}(\ell_j)\right] = \tan\left[\varphi^{(+)}(-\ell_j)\right] \tag{11}$$

that is

$$\varphi^{(-)}(\ell_j) = \varphi^{(+)}(-\ell_j) + \pi, \tag{12}$$

The two orthogonal polarizations are sorted over two overlapping heptagons, as shown in the scheme in Figure 3b. According to Equation (12), the far-field intensity peaks are expected to be at the following angular positions for the left-handed beams:

$$\varphi_j^{(-)} = (7 - j + 1)\frac{2\pi}{7} + \pi = -(j-1)\frac{2\pi}{7} + \pi, \tag{13}$$

As shown in Figure 3b, for increasing OAM values, the corresponding spots appear counterclockwise (clockwise) for incident right-handed (left-handed) circular polarization.

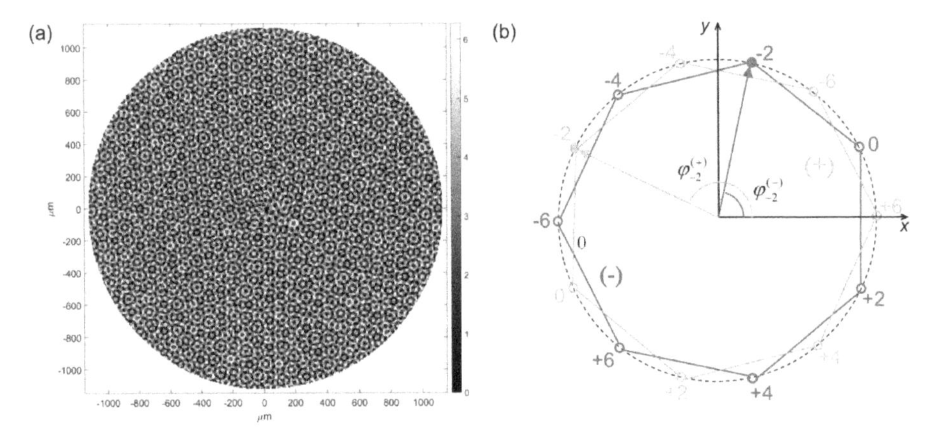

Figure 3. (**a**) Numerical phase pattern for the demultiplexing of optical beams with OAM in the set {−6, −4, −2, 0, +2, +4, +6} on a heptagonal configuration. Pixel size: 6.250 μm × 6.250 μm. 16 phase levels. Radius length: 180 pixels. (**b**) Far-field channel scheme for the given OAM set and circular polarization states. Right-handed (in blue) and left-handed (in red) circularly-polarized OAM beams are detected in far field on two distinct heptagons.

2.1.3. Equally-Spaced Hemi-Circular Configuration

In order to arrange the far-field channel at equally-spaced angular positions without overlap, a semicircle configuration appears to be the best choice. In this case we considered the set {−9, −6, −3, 0, +3, +6, +9} and we fixed the carrier spatial frequencies in such a way that the far-field peaks were arranged over a semicircle of constant radius r at equally-spaced angular positions (see Figure 4b), specified as follows:

$$\begin{aligned} r_j^{(+)} &= r = \gamma \tfrac{f}{k} \\ \varphi_j^{(+)} &= j\tfrac{2\pi}{2n} = j\tfrac{2\pi}{14} \end{aligned}, \tag{14}$$

where $j = 1, \ldots 7$. According to Equation (7), the far-field intensity peaks are expected to be at the following angular positions:

$$\varphi_j^{(-)} = (7 - j + 1)\frac{2\pi}{14} + \pi = -(j-1)\frac{2\pi}{14}, \tag{15}$$

The two orthogonal polarizations are therefore sorted over two complementary semicircles without overlapping, as shown in the scheme in Figure 4b.

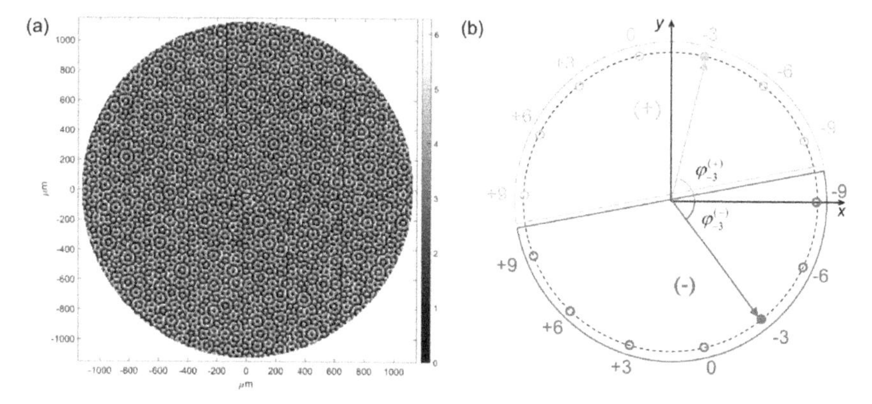

Figure 4. (**a**) Numerical phase pattern for the demultiplexing of optical beams with OAM in the set {−9, −6, −3, 0, +3, +6, +9} on a semicircular configuration. Pixel size: 6.250 μm × 6.250 μm. 16 phase levels. Radius length: 180 pixels. (**b**) Far-field channel scheme for the given OAM set and circular polarization states. Right-handed (in blue) and left-handed (in red) circularly-polarized OAM beams are detected in far field on two distinct and complementary semicircles.

2.2. Subwavelength Grating Design

The metasurface version of the computed optical elements has been realized in the form of spatially-variant subwavelength gratings, whose ridges orientation is rotated pixel-by-pixel introducing a spatially-dependent form birefringence. The key element of the metasurface is represented by the subwavelength linear grating cell, whose local orientation $\theta(u,v)$ is fixed in order to transfer the desired geometric-phase $\Omega(u,v)$ to the input wavefront, according to Reference [36]:

$$\theta(u,v) = \frac{\Omega(u,v)}{2} \tag{16}$$

being (u,v) the coordinates of the reference frame on the optical element plane. The phase-patterns of the designed optical elements have been calculated as 4-bit grayscale images (16 phase levels) and converted into subwavelength grating metasurfaces with custom MATLAB® codes. The gray level j, in the range from 0 to 15, has been associated to the rotation angle $j2\pi/32$ of the corresponding subwavelength grating vector. For a given grating thickness, numerical simulations must be performed in order to identify the optimal profile, in terms of duty-cycle and period, providing the maximum conversion efficiency, i.e., π-delay between ordinary and extraordinary axes. In Reference [36], a numerical study was performed implementing Rigorous Coupled-Wave Analysis (RCWA) [37,38] for a binary silicon grating in air at 1310 nm, in order to extract the optimal configurations of period and duty-cycle which provide π-retardation. For a thickness of 535 nm with a duty-cycle around 0.5, the grating period providing a π-delay is around 290 nm. This configuration was chosen for the design and fabrication of the silicon PBOEs presented in this study.

2.3. Fabrication

For the fabrication of subwavelength gratings with high aspect ratio a three-step stamp process was considered. Electron-beam lithography (EBL) provides the ideal method to transfer the computational patterns from a digitally-stored format to a physical layer with high-resolution profiles. The original EBL pattern was transformed into an imprinting mold for subsequent imprinting replica and inductively coupled plasma—reactive ion etching (ICP-RIE) to achieve the final sample.

Electron-beam lithography was performed with a JBX-6300FS EBL machine (JEOL, Tokyo, Japan) 12 MHz, 5 nm resolution, working at 100 kV with a current of 100 pA. A thin layer of positive resist (AR-P 672.03, ALLRESIST GmbH, Strausberg, Germany) was spun at 4000 rpm obtaining a thickness around 130 nm, followed by a hot plate soft-baking process at 150 °C for 3 min. Afterwards, the sample was developed in an isopropyl alcohol (IPA):deionized water 7:3 solution for 60 s, in order to remove the exposed areas.

To achieve the transfer from the EBL-patterned resist to the Silicon substrate, a 7-seconds stripping process in O_2 plasma was performed, followed by a 72-seconds etching in fluorine-based plasma with STS MESC MULTIPLEX ICP (SemiStar Corp, Morgan Hill, CA, USA).

Next, a Thermal-NanoImprint Lithography (T-NIL) was performed with a Paul-Otto Weber hydraulic press with heating/cooling plates, for high-resolution replica [39,40]. The process was conducted using the previously-etched EBL master as cast after a silanization process with Trichloro(1H,1H,2H,2H-perfluorooctyl)silane PFOTS (Thermo Fisher (Kandel) GmbH, Karlsruhe, Germany) [41,42]. A layer of MR-I 7010E was deposited on a silicon wafer at 1750 rpm, achieving a thickness around 120 nm, followed by a 2-min soft bake at 140 °C. The sample was placed in contact with the master within a system of compliances in order to homogenize the temperature and pressure on the entire surface. The T-NIL process was performed at 100 °C for 10 min at 100-bar pressure. At the end of the imprinting step, a temperature decrease down to 35 °C occurred, maintaining the pressure fixed at 100 bar.

After a 13-second O_2 treatment to remove the residual layer, a 10-nm Cr hard mask was deposited by *e*-gun evaporation and the transfer of the resist pattern was carried out by a lift-off process in a sonicated acetone bath for 180 s. Finally, an ICP-RIE etching was performed to remove the residual layer and hence obtain the required grating thickness. The etching time was finely adjusted to reach a final depth around 535 nm, as recommended by numerical simulations. In Figure 5, inspections at scanning electron microscopy (SEM) of the final sample are shown. In particular, the well-defined line profile is evidence of the suitability of the nanofabrication recipe for pattern transfer onto the silicon substrate.

Figure 5. (**a**) SEM inspections of the fabricated PBOE on silicon substrate performing PDM and OAM-MDM according to the scheme in Figure 3 (heptagonal configuration). (**b–d**) Details at higher magnification. Grating period $\Lambda = 290$ nm, duty-cycle 0.5, thickness 535 nm, pixel size 6.250 μm. 16 rotation angles.

2.4. Optical Characterization Setup

The experimental setup for the optical analysis of the fabricated samples is depicted in Figure 6. The performance of the metasurfaces has been analyzed with input optical beams endowed with integer orbital angular momentum, generated by uploading the proper phase patterns on a LCoS spatial light modulator (SLM) (X13267-08, Hamamatsu, Shizuoka, Japan) with amplitude/phase modulation [43]. An aspheric lens with focal length $f_F = 7.5$ mm (A375TM-C, Thorlabs, Newton, NJ, USA) was used to collimate the output of a DFB laser ($\lambda = 1310$ nm) emerging at the end of a single mode fiber. Then the output beam was linearly polarized and expanded with a first telescope ($f_1 = 3.5$ cm, $f_2 = 10.0$ cm) before illuminating the display of the SLM. A beam-splitter (50:50) was inserted after the telescope in order to produce a second coherent Gaussian beam for interferometric analysis. A second telescope ($f_3 = 20.0$ cm, $f_4 = 12.5$ cm) with an aperture in the Fourier plane was used to isolate and image the first-order encoded mode onto the sorter. A second beam-splitter (50:50) was used to split the beam and check the input beam profile with a first camera (WiDy SWIR 640U-S, NIT, Verrières-le-Buisson, France). A Mach–Zehnder interferometric bench was added, as shown in Figure 6, in order to analyze the phase pattern of the modes generated with the SLM. Afterwards, the OAM beam illuminated the silicon sample, mounted on a 6-axis kinematic mount with micrometer drives (K6XS, Thorlabs, Newton, NJ, USA). Finally, a second camera (WiDy SWIR 640U-S, NIT, Verrières-le-Buisson, France) was used to collect the far field at the back-focal plane of a lens with focal length $f_5 = 7.5$ cm. A sequence of linear polarizer (LPIREA100-C, Thorlabs, Newton, NJ, USA) and quarter-wave plate (WPQ10M-1310, Thorlabs, Newton, NJ, USA) was placed before and after the sorter, in reverse order, to control and select the circular polarization state of the input and output beams.

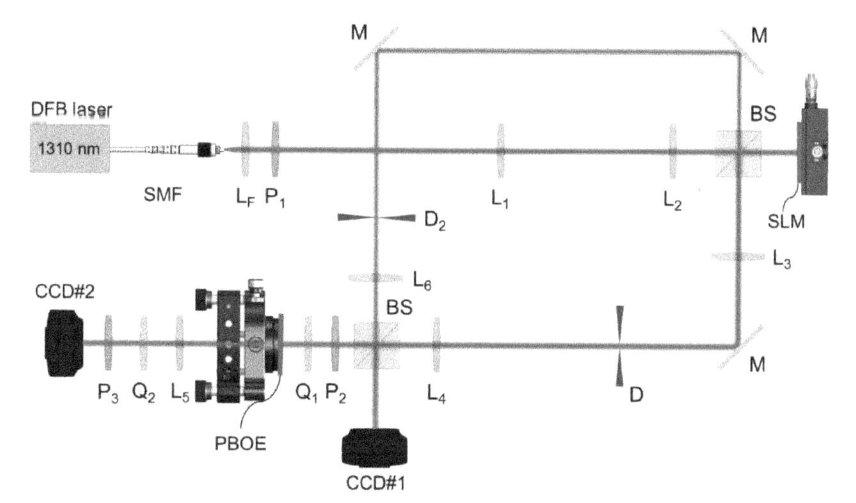

Figure 6. Experimental setup used for the optical analysis of the fabricated Pancharatnam–Berry optical elements (PBOE). The output of the DFB laser ($\lambda = 1310$ nm) is collimated after a single mode fiber (SMF) using an aspheric lens (focal length $f_F = 7.5$ mm), linearly polarized (P_1) and magnified with a first telescope ($f_1 = 3.5$ cm, $f_2 = 10.0$ cm). The first order of the spatial light modulator (SLM) used for OAM-beam generation is filtered (D) and resized ($f_3 = 20.0$ cm, $f_4 = 12.5$ cm) before impinging on the demultiplexer. A beam splitter (BS) is exploited both to check the input beam and collect the output intensity at the back-focal plane of a fifth lens ($f_5 = 7.5$ cm). A sequence of quarter-wave plates (Q) and linear polarizers (P) is placed before (P_2, Q_1) and after (Q_2, P_3) the sorter, in reverse order, in order to control and select the desired circular polarization. A Mach–Zehnder interferometric setup is used to analyze the spiralgram of the input optical vortices and infer the carried OAM value and sign.

3. Results

The output of the fabricated PBOEs has been analyzed and recorded for input circularly-polarized beams with well-defined OAM. For each PBOE, beams carrying OAM in the sorting set of the selected metasurface have been produced, in sequence, and circularly polarized before impinging on the optical element, according to the scheme in Figure 6. Using a Mach–Zehnder interferometric bench, as shown in Figure 6, the interference pattern between the generated OAM beam and a reference Gaussian beam was generated and collected in order to check the input OAM value. As a matter of fact, since the phase structure of an integer-OAM beam presents ℓ intertwined helical phase fronts, being ℓ the amount of OAM, the interference with a coaxial Gaussian beam generates a fringe pattern of ℓ spirals, whose helicity is given by the sign of ℓ [44] (Figure 7a, Figure 8a, and Figure 9a).

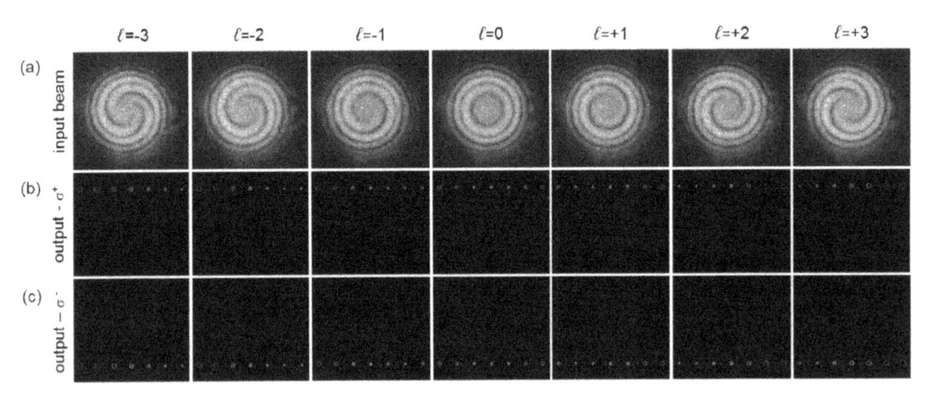

Figure 7. Optical characterization of the demultiplexer in Figure 2. (**a**) Experimental interference pattern of the input beams. The twist-handedness and number of the spiral arms reveal the sign and value of orbital angular momentum, respectively. Experimental output intensity for input right-handed (**b**) and left-handed (**c**) circular polarization states. The position of the far-field bright spots is in accordance with the channel scheme depicted in Figure 2b.

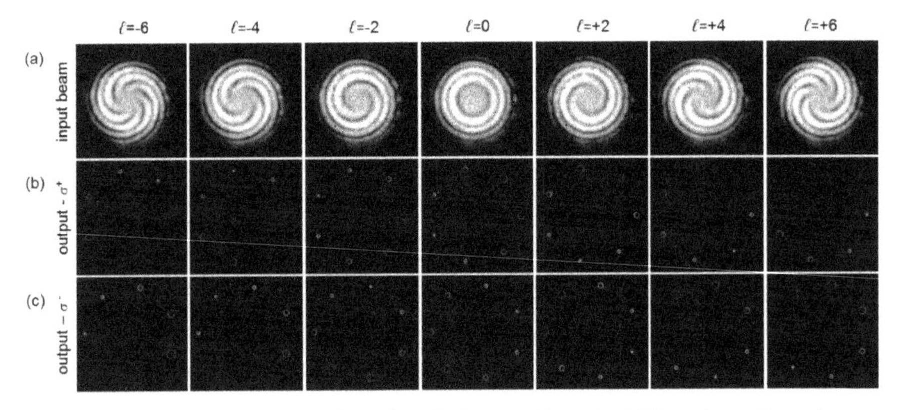

Figure 8. Optical characterization of the demultiplexer in Figure 3. (**a**) Experimental interference pattern of the input beams. The twist-handedness and number of the spiral arms reveal the sign and value of orbital angular momentum, respectively. Experimental output intensity for input right-handed (**b**) and left-handed (**c**) circular polarization states. The position of the far-field bright spots is in accordance with the channel scheme depicted in Figure 3b.

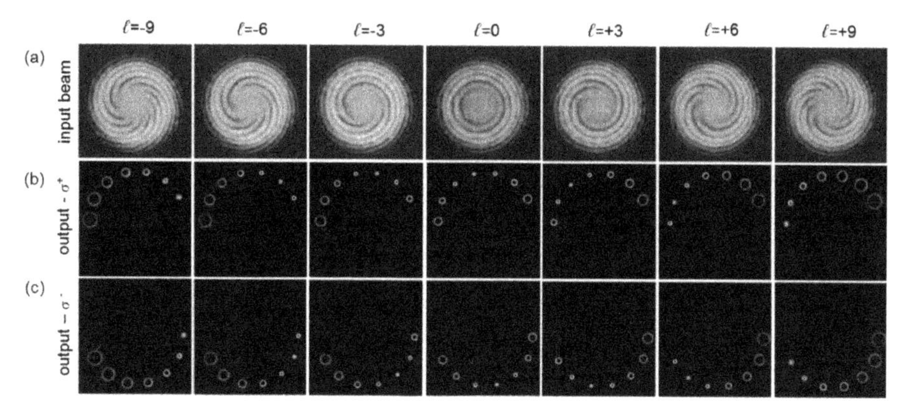

Figure 9. Optical characterization of the demultiplexer in Figure 4. (**a**) Experimental interference pattern of the input beams. The twist-handedness and number of the spiral arms reveal the sign and value of orbital angular momentum, respectively. Experimental output intensity for input right-handed (**b**) and left-handed (**c**) circular polarization states. The position of the far-field bright spots is in accordance with the channel scheme depicted in Figure 4b.

When a beam carrying OAM illuminates the demultiplexer, the optical element performs the projection over the mode set for which the phase pattern has been calculated. Next, a bright spot is detected in correspondence of the input OAM value, when it is present, at the coordinates given by corresponding far-field channel scheme. Otherwise, a non-null OAM beam is generated, i.e., an annular intensity profile with a central dark singularity.

In Figure 7, the optical characterization is reported for the PBOE performing OAM sorting in the range $\{-3, \ldots, +3\}$ with OAM step $\Delta\ell = 1$ (Figure 7a). In Figure 7b,c, the far field is shown for input beams with right-handed and left-handed circular polarizations, respectively. As expected, the demultiplexer can sort the orthogonal polarization states onto two different linear arrays, while the OAM value is detected correctly according to the scheme in Figure 2b. A similar analysis is reported in Figure 8, for the PBOE performing OAM demultiplexing in the range $\{-6, \ldots, +6\}$, step $\Delta\ell = 2$, according to the scheme in Figure 3b. In Figure 8b,c, the far field is shown for input optical vortices with right-handed and left-handed circular polarizations, respectively. The demultiplexer separates the orthogonal polarization states onto two heptagons, while the OAM content is detected correctly. Figure 9 reports the optical analysis of the PBOE performing sorting in the OAM range $\{-9, \ldots, +9\}$, step $\Delta\ell = 3$, over a circular configuration, as depicted in the scheme in Figure 4b. In Figure 9b,c, the far field is shown for input vortices with right-handed and left-handed circular polarization, respectively. As expected, the demultiplexer can distinguish between orthogonal polarization states, by projecting them onto two complementary, i.e., non-overlapping, semicircles.

4. Discussion

In this work, we described the design, nanofabrication, and optical characterization of silicon metasurfaces for the parallel sorting of orbital angular momentum and polarization using the method of optical-mode projection. The samples were fabricated in the form of dielectric Pancharatnam–Berry optics, whose inhomogeneous anisotropy imparts a spatially-variant phase-change due to a local control of the input polarization. In particular, the phase is geometric in nature and equal to twice the rotation angle of the local extraordinary axis, corresponding to the direction of the subwavelength grating vector. Three different sorters have been designed and fabricated, performing combined PDM and OAM-MDM over different OAM sets and channel configurations, with the aim to exhibit the versatility of the demultiplexing method in terms of channels geometry and OAM values. In particular, we demonstrated the possibility to sort a symmetric range of OAM beams over a linear array and over 2D regular distributions, specifically a regular polygon and a semicircle. By properly designing the far-field channel scheme, fixed by the spatial frequency carriers in the phase pattern definition of the sorter, it was possible to originate two non-overlapping channel geometries for the two orthogonal polarizations. The optical characterization has been reported at the wavelength of 1310 nm, in the telecom O-band, showing the expected capability to distinguish between modes with different orbital angular momentum and spin values by using a single element. With respect to the diffractive counterpart [23], the number of available channels is redoubled without the need of additional optical elements.

Metasurfaces have become one of the most rapidly expanding frontiers of nanophotonics to revolutionize optics by substituting refractive and diffractive optics in many widespread applications and introducing entirely altogether novel functionalities [45,46]. In particular, the possibility to use silicon as optical material has promoted the flourishing of a new framework in which optics design and silicon photonics merge to create a new generation of optical elements with unprecedented levels of integration. In comparison with plasmonic metamaterials, the importance of silicon in optics design and fabrication is based not only on its optical properties, low-cost, and well-established nanofabrication techniques, but also on the peculiar and promising prospects that silicon nanostructures can provide in terms of integration into existing photonic architectures and complementary metal-oxide semiconductor (CMOS) compatibility [47,48].

By including optics design and silicon photonics, the presented metasurfaces pave the way to novel optical devices for combined polarization- and OAM-mode division multiplexing with an unprecedented combination of miniaturization and integration.

Author Contributions: Conceptualization, G.R. and F.R.; formal analysis, G.R.; investigation, G.R., M.M., and P.C.; software, G.R.; methodology, M.M., P.C.; writing—original draft preparation, G.R., P.C., and M.M.; writing—review and editing, F.R.; supervision, F.R.; funding acquisition, F.R.

Funding: This work was supported by 3M Optics S.r.l.—SIAF Group, and by CEPOLISPE project 'VORTEX 2'.

Acknowledgments: The authors gratefully thank Giuseppe Parisi and Ing. Mauro Zontini for the interesting discussions during this work.

Conflicts of Interest: The authors declare no conflict of interest.

References

1. Ritsch-Marte, M. Orbital angular momentum light in microscopy. *Philos. Trans. R. Soc. A* **2017**, *375*, 20150437. [CrossRef] [PubMed]
2. Vicidomini, G.; Bianchini, P.; Diaspro, A. STED super-resolved microscopy. *Nat. Methods* **2018**, *15*, 173–182. [CrossRef] [PubMed]
3. Mari, E.; Tamburini, F.; Swartzlander, G.A.; Bianchini, A.; Barbieri, C.; Romanato, F.; Thidé, B. Sub-Rayleigh optical vortex coronagraphy. *Opt. Express* **2012**, *20*, 2445–2451. [CrossRef] [PubMed]
4. Padgett, M.J.; Bowman, R. Tweezers with a twist. *Nat. Photonics* **2011**, *5*, 343–348. [CrossRef]
5. Ruffato, G.; Rossi, R.; Massari, M.; Mafakheri, E.; Capaldo, P.; Romanato, F. Design, fabrication and characterization of Computer-Generated Holograms for anti-counterfeiting applications using OAM beams as light decoders. *Sci. Rep.* **2017**, *7*, 18011. [CrossRef] [PubMed]
6. Wang, J. Twisted optical communications using orbital angular momentum. *China Phys. Mech. Astron.* **2019**, *62*, 34201. [CrossRef]
7. Mirhosseini, M.; Magana-Loaiza, O.S.; O'Sullivan, M.N.; Rudenburg, B.; Malik, M.; Lavery, M.P.J.; Padgett, M.J.; Gauthier, D.J.; Boyd, R.W. High-dimensional quantum cryptography with twisted light. *New J. Phys.* **2015**, *17*, 033033. [CrossRef]
8. Allen, L.; Beijersbergen, M.W.; Spreeuw, R.J.C.; Woerdman, J.P. Orbital angular momentum of light and the transformation of Laguerre-Gaussian modes. *Phys. Rev. A* **1992**, *45*, 8185–8189. [CrossRef] [PubMed]
9. Padgett, M.J. Orbital angular momentum 25 years on. *Opt. Express* **2017**, *25*, 11265–11274. [CrossRef] [PubMed]

10. Agrell, E.; Karlsson, M.; Chraplyvy, A.R.; Richardson, D.J.; Krummrich, P.M.; Winzer, P.; Roberts, K.; Fisher, J.K.; Savory, S.J.; Eggleton, B.J.; et al. Roadmap of optical communications. *J. Opt.* **2016**, *18*, 063002. [CrossRef]
11. Andrews, D.; Babiker, M. *The Angular Momentum of Light*; Cambridge University Press: Cambridge, UK, 2013; ISBN 9781107006348.
12. Yu, S. Potential and challenges of using orbital angular momentum communications in optical interconnects. *Opt. Express* **2015**, *23*, 3075–3087. [CrossRef] [PubMed]
13. Willner, A.E.; Ren, Y.; Xie, G.; Yan, Y.; Li, L.; Zhao, Z.; Wang, J.; Tur, M.; Molish, A.F.; Ashrafi, S. Recent advances in high-capacity free-space optical and radio-frequency communications using orbital angular momentum multiplexing. *Philos. Trans. A Math Phys. Eng. Sci.* **2017**, *375*, 20150439. [CrossRef]
14. Bozinovic, N.; Yue, Y.; Ren, Y.; Tur, N.; Kristensen, P.; Huang, H.; Willner, A.E.; Ramachandran, S. Terabit-scale orbital angular momentum mode division multiplexing in fibers. *Science* **2013**, *340*, 1545–1548. [CrossRef] [PubMed]
15. Ramachandran, S.; Kristensen, P. Optical vortices in fiber. *Nanophotonics* **2013**, *2*, 455–474. [CrossRef]
16. Winzer, P.J.; Neilson, D.T.; Chraplyvy, A.R. Fiber-optic transmission and networking: The previous 20 and the next 20 years. *Opt. Express* **2018**, *26*, 24190–24239. [CrossRef] [PubMed]
17. Sit, A.; Bouchard, F.; Fickler, R.; Cagnon-Bischoff, J.; Larocque, H.; Heshami, K.; Elser, D.; Peuntinger, C.; Gunthner, K.; Heim, B.; et al. High-dimensional intracity quantum cryptography with structured photons. *Optica* **2017**, *4*, 1006–1010. [CrossRef]
18. Bouchard, F.; Heshami, K.; England, D.; Fickler, R.; Boyd, R.W.; Englert, B.-G.; Sanchez-Soto, L.L.; Karimi, E. Experimental investigation of high-dimensional quantum key distribution protocols with twisted photons. *Quantum* **2018**, *2*, 111. [CrossRef]
19. Qu, Z.; Djordjevic, I.B. High-speed free-space optical continuous variable quantum key distribution enabled by three-dimensional multiplexing. *Opt. Express* **2017**, *25*, 7919–7928. [CrossRef]
20. Larocque, H.; Gagnon-Bischoff, J.; Mortimer, D.; Zhang, Y.; Bouchard, F.; Upham, J.; Grillo, V.; Boyd, R.W.; Karimi, E. Generalized optical angular momentum sorter and its application to high-dimensional quantum cryptography. *Opt. Express* **2017**, *25*, 19832–19843. [CrossRef]
21. Ndagano, B.; Nape, I.; Perez-Garcia, B.; Scholes, S.; Hernandez-Aranda, R.I.; Konrad, T.; Lavery, M.P.J.; Forbes, A. A deterministic detector for vector vortex states. *Sci. Rep.* **2017**, *7*, 13882. [CrossRef]
22. Wan, C.; Rui, G.; Chen, J.; Zhan, Q. Detection of photonic orbital angular momentum with micro- and nano-optical structures. *Front. Optoelectron.* **2017**, *12*, 88–96. [CrossRef]
23. Ruffato, G.; Massari, M.; Romanato, F. Diffractive optics for combined spatial- and mode- division demultiplexing of optical vortices: Design, fabrication and optical characterization. *Sci. Rep.* **2016**, *6*, 24760. [CrossRef] [PubMed]
24. Ruffato, G.; Massari, M.; Romanato, F. Compact sorting of optical vortices by means of diffractive transformation optics. *Opt. Lett.* **2017**, *42*, 551–554. [CrossRef] [PubMed]
25. Ruffato, G.; Massari, M.; Parisi, G.; Romanato, F. Test of mode-division multiplexing and demultiplexing in free-space with diffractive transformation optics. *Opt. Express* **2017**, *25*, 7859–7868. [CrossRef] [PubMed]
26. Ruffato, G.; Girardi, M.; Massari, M.; Mafakheri, E.; Sephton, B.; Capaldo, P.; Forbes, A.; Romanato, F. A compact diffractive sorter for high-resolution demultiplexing of orbital angular momentum beams. *Sci. Rep.* **2018**, *8*, 10248. [CrossRef] [PubMed]
27. Berkhout, G.C.G.; Lavery, M.P.J.; Courtial, J.; Beijersbergen, M.W.; Padgett, M.J. Efficient sorting of orbital angular momentum states of light. *Phys. Rev. Lett.* **2010**, *105*, 153601. [CrossRef] [PubMed]
28. Gibson, G.; Courtial, J.; Padgett, M.J.; Vasnetsov, M.; Pas'ko, V.; Barnett, S.M.; Franke-Arnold, S. Free-space information transfer using light beams carrying orbital angular momentum. *Opt. Express* **2004**, *12*, 5448–5456. [CrossRef] [PubMed]
29. Trichili, A.; Rosalez-Guzman, C.; Dudley, A.; Ndagano, B.; Salem, A.B.; Zghal, M.; Forbes, A. Optical communication beyond orbital angular momentum. *Sci. Rep.* **2016**, *6*, 27674. [CrossRef] [PubMed]
30. Roux, F.S. Geometric phase lens. *J. Opt. Soc. Am. A* **2006**, *23*, 476–482. [CrossRef]
31. Chen, M.L.N.; Jiang, L.J.; Sha, W.E.I. Orbital Angular Momentum Generation and Detection by Geometric-Phase Based Metasurfaces. *Appl. Sci.* **2018**, *8*, 362. [CrossRef]
32. Desiatov, B.; Mazurski, N.; Fainman, Y.; Levy, U. Polarization selective beam shaping using nanoscale dielectric metasurfaces. *Opt. Express* **2015**, *23*, 22611–22618. [CrossRef] [PubMed]

33. Li, Y.; Li, X.; Chen, L.; Pu, M.; Jin, J.; Hong, M.; Luo, X. Orbital Angular Momentum Multiplexing and Demultiplexing by a Single Metasurface. *Adv. Opt. Mater.* **2017**, *5*, 1600502. [CrossRef]
34. Emoto, A.; Nishi, M.; Okada, M.; Manabe, S.; Matsui, S.; Kawatsuki, N.; Ono, H. Form birefringence in intrinsic birefringent media possessing a subwavelength structure. *Appl. Opt.* **2010**, *49*, 4355–4361. [CrossRef] [PubMed]
35. Kotlyar, V.V.; Khonina, S.N.; Soifer, V.A. Light field decomposition in angular harmonics by means of diffractive optics. *J. Mod. Opt.* **1998**, *45*, 1495–1506. [CrossRef]
36. Capaldo, P.; Mezzadrelli, A.; Pozzato, A.; Ruffato, G.; Massari, M.; Romanato, F. Nano-fabrication and characterization of silicon meta-surfaces provided with Pancharatnam-Berry effect. *Opt. Mater. Express* **2019**, *9*, 1015–1032. [CrossRef]
37. Moharam, M.G.; Pommet, D.A.; Grann, E.B.; Gaylord, T.K. Stable implementation of the rigorous coupled-wave analysis for surface-relief gratings: Enhanced transmittance matrix approach. *J. Opt. Soc. Am. A* **1995**, *12*, 1077–1086. [CrossRef]
38. Kikuta, H.; Ohira, Y.; Kubo, H.; Iwata, K. Effective medium theory of two-dimensional subwavelength gratings in the non-quasi-static limit. *J. Opt. Soc. Am. A* **1998**, *15*, 1577–1585. [CrossRef]
39. Beck, M.; Graczyk, M.; Maximov, I.; Sarwe, E.L.; Ling, T.G.I.; Keil, M.; Montelius, L. Improving stamps for 10 nm level wafer scale nanoimprint lithography. *Microelectron. Eng.* **2002**, *61–62*, 441–448. [CrossRef]
40. Pozzato, A.; Grenci, G.; Birarda, G.; Tormen, M. Evaluation of a novolak based positive tone photoresist as NanoImprint Lithography resist. *Microelectron. Eng.* **2011**, *88*, 2096–2099. [CrossRef]
41. DePalma, V.; Tillman, N. Friction and Wear of Self-Assembled Trichlorosilane Monolayer Films on Silicon. *Langmuir* **1989**, *5*, 868–872. [CrossRef]
42. Haensch, C.; Hoeppener, S.; Schubert, U.S. Chemical modification of self-assembled silane based monolayers by surface reactions. *Chem. Soc. Rev.* **2010**, *39*, 2323–2334. [CrossRef] [PubMed]
43. Rosales-Guzmán, C.; Forbes, A. *How to Shape Light with Spatial Light Modulators*; SPIE Press: Bellingham, WA, USA, 2017; ISBN 9781510613027.
44. Padgett, M.; Courtial, J.; Allen, L. Light's Orbital Angular Momentum. *Phys. Today* **2004**, *57*, 35–40. [CrossRef]
45. Koenderink, A.F.; Alù, A.; Polman, A. Nanophotonics: Shrinking light based technology. *Science* **2015**, *348*, 516. [CrossRef] [PubMed]
46. Capasso, F. The future and promise of flat optics: A personal perspective. *Nanophotonics* **2018**, *7*, 953–957. [CrossRef]
47. Jahani, S.; Jacob, Z. All-dielectric metamaterials. *Nat. Nanotechnol.* **2016**, *11*, 23–36. [CrossRef]
48. Staude, I.; Schilling, J. Metamaterial-inspired silicon nanophotonics. *Nat. Photonics* **2017**, *11*, 274–284. [CrossRef]

Article

Orbital Angular Momentum Multiplexed Free-Space Optical Communication Systems Based on Coded Modulation

Zhen Qu * and Ivan B. Djordjevic

Department of Electrical and Computer Engineering, University of Arizona, 1230 E. Speedway Blvd., Tucson, AZ 85721, USA; ivan@email.arizona.edu
* Correspondence: zhenqu@email.arizona.edu; Tel.: +1-520-442-7197

Received: 22 October 2018; Accepted: 4 November 2018; Published: 7 November 2018

Abstract: In this paper, we experimentally investigate the turbulence mitigation methods in free-space optical communication systems based on orbital angular momentum (OAM) multiplexing. To study the outdoor atmospheric turbulence environment, we use an indoor turbulence emulator. Adaptive optics, channel coding, Huffman coding combined with low-density parity-check (LDPC) coding, and spatial offset are used for turbulence mitigation; while OAM multiplexing and wavelength-division multiplexing (WDM) are applied to boost channel capacity.

Keywords: free-space optical communications; orbital angular momentum; turbulence mitigation

1. Introduction

Optical communication systems are usually deployed over fiber-optic links [1–3], free-space optical (FSO) links [4–6], or hybrid FSO fiber links [7]. Recent advances in photonics integrated circuits (PIC) have been greatly pushing forward the worldwide application of optical communications [8–10]. Although they have enabled a capacity-approaching communication [11–13], fiber-optic links may be too fragile or costly to be deployed in some environments, e.g., seismic belts. As a result, FSO links are more favorable due to their easy and fast communication link reconstruction. Although the channel loss of FSO links is not stable, and typically higher than that of fiber-optic links, FSO communication systems provide free-scalable channels for spatial mode division (SDM) multiplexing, e.g., orbital angular momentum (OAM) multiplexing [14–17]. Despite their free-scalable characteristics, spatially multiplexed modes hardly preserve their orthogonality when transmitting over the atmospheric FSO links, resulting in dynamic inter-mode crosstalk [18–20].

There are several ways to mitigate the inter-mode crosstalk, including wavefront sensor (WFS) or wavefront sensorless based adaptive optics (AO) systems [21–23], digital multi-input multi-output (MIMO) equalization [24–26], and advanced forward error correction (FEC) based channel coding techniques [27–29]. AO systems are usually implemented in satellite communications to mitigate wavefront distortion, while their commercial application in near-Earth FSO links is greatly limited by the expensive WFS and deformable mirror (DM). In some instances, a wavefront sensorless AO is used as a trade-off between cost and reliability. MIMO equalization is also often used to relieve inter-mode crosstalk among the multiplexed spatial modes. It is, however, preferentially used in the FSO links affected by weak-to-medium atmospheric turbulence. When spatial modes are transmitted in a strong atmospheric turbulence environment, the unwanted inter-mode crosstalk is not only limited to adjacent spatial modes, but also spread widely across other spatial modes. As a result, the use of more mode detectors is expected at the receiver side to capture the distorted signals in correlated modes, followed by more computationally complex MIMO equalization. If the quantity of mode detector is not sufficient, MIMO equalization may fail to work due to data loss. FEC based channel coding techniques have been fully developed over decades, and extensively used in error-free digital fiber-optic communication systems. However, FEC coding-only solution can't guarantee a reliable data transmission over long reach FSO links, especially in the strong atmospheric turbulence environment.

In this paper, we discuss the high-speed OAM multiplexed FSO communication systems, enabled by AO, low-density parity-check (LDPC) coding, spatial offset (SO), and joint Huffman and LDPC coding. First, we experimentally study an AO assisted, LDPC coded, OAM-based FSO communication system. Briefly, four OAM multiplexed mode channels that in total carry 500 Gbps quadrature phase shift keying (QPSK) signals are transmitted over wavelength-division multiplexed (WDM) channels with 50 GHz spacing. The turbulence-induced inter-mode crosstalk is compensated by a wavefront sensorless AO setup. Subsequently, error-free communication can be achieved with a strong LDPC coding scheme. The minimum coding gain of 3.9 dB is achieved at BER = 2×10^{-2} for OAM states ± 2 and ± 6.

Second, we present another inter-mode crosstalk mitigation solution in an atmospheric turbulence limited OAM multiplexed FSO links. The OAM mode crosstalk is first relieved by optical spatial mode offset, and then further resolved by coded modulation technology. Huffman coding and optimal constellation design techniques are applied to generate the quadrature amplitude modulation (QAM) formats, i.e., 5/9-QAM formats. Meanwhile, the GF(5) based LDPC coding is implemented for 5-QAM symbol sequences, and GF(32) based LDPC coding is implemented for 9-QAM symbol sequences [30,31]. Unlike the classical OAM multiplexed FSO links, where all spatial mode channels are centrally aligned, binary FEC coding is used for error correction. Furthermore, uniformly distributed M-QAM formats are used for data modulation, e.g., QPSK and 8-QAM. The proposed two-stage OAM mode crosstalk mitigation solution can largely enhance the communication reliability in atmospheric FSO links.

2. Adaptive Optics Enabled Free-Space Optical Communication

2.1. Experimental Setup

Figure 1 shows the experimental setup for an AO enabled FSO communication system. The five continuous wave (CW) laser beams are generated with the inter-channel spacing of 50 GHz (1549.32–1550.92 nm). The wavelength channels are multiplexed together and used as the optical input of an I/Q modulator. The pseudorandom binary sequence (PRBS) signals are encoded using a binary LDPC code with the code rate of 0.8. The data streams pass to the arbitrary waveform generator (AWG) and drive the I/Q modulator to generate a 15.6 GBaud optical QPSK signal. An optical interleaver (IL) is applied to separate odd and even channel signals, which are then decorrelated by 350 symbols and recombined together. The resulting WDM QPSK optical signals with the decorrelated adjacent wavelength channels are boosted by an Erbium-doped fiber amplifier (EDFA), followed by an optical tunable filter (OTF) to suppress amplified spontaneous emission (ASE) noise. The optical signals are separated by a coupler, and one path is later decorrelated before recombination. The optical Gaussian modes are collimated by fixed fiber optic collimator and converted to OAM modes (OAM states ± 2 and ± 6) by using a high-resolution spatial light modulator (SLM). The resulting OAM modes are then centrally aligned by a beam splitter (BS). Another 1548.9 nm Gaussian probe beam is used to assist AO compensation. The Gaussian probe beam is expanded using a beam expander (BE) to reach a diameter larger than the widest OAM beams (OAM states ± 6) generated in this experiment. At this stage, the probe beam is also centrally aligned with the data-carrying OAM modes.

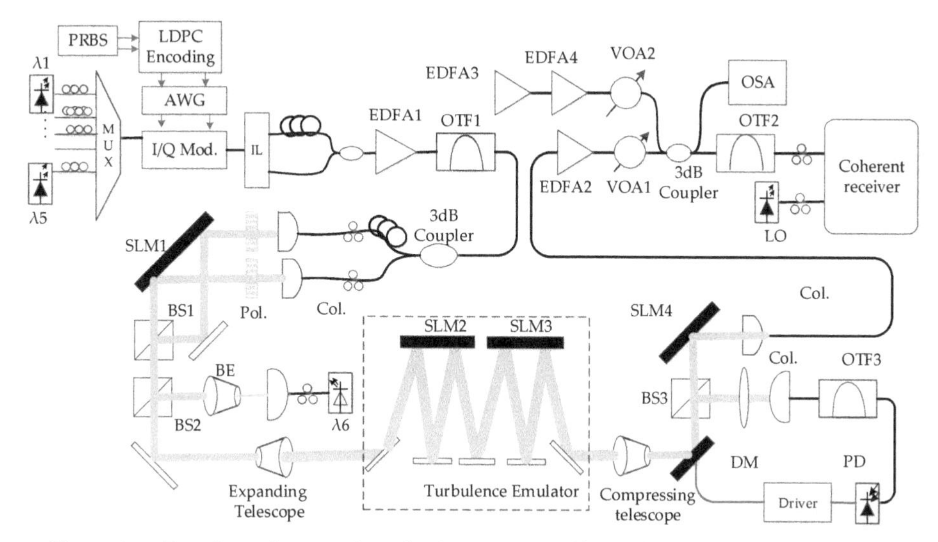

Figure 1. Experimental setup for adaptive optics (AO) enabled free-space optical (FSO) communication system.

The expanding telescope is applied to adjust the diameters of the collimated OAM beam and Gaussian probe beam, which are then sent to the turbulence emulator. The SLM2 and SLM3 are continuously and randomly updating phase patterns to be modelled on the dynamic atmospheric turbulence environment [22]. The accuracy of the turbulence model can be validated in terms of on-axis Gamma–Gamma intensity distribution and intensity correlation function. The atmospheric turbulence emulator used here is designed according to the Rytov variance [32] of $\sigma_R^2 = 2$.

The size of the distorted OAM and probe beams are decreased by a compressing telescope. The distorted beams then pass to the DM for wavefront correction. The distorted Gaussian probe beam functions as a stimulus in this wavefront sensorless AO setup, with the assumption that the probe beam and OAM beams are affected by the similar wavefront distortion. Partial optical beams are segregated via a BS and collected by a single-mode fiber (SMF) patch cable. In this experiment, the OTF3 with a central wavelength of 1548.9 nm is implemented to only capture the Gaussian probe beam, followed by a photodiode (PD) for power monitoring. The detected analog voltage is digitized by an analog-to-digital converter (ADC) to update the DM pixels based on stochastic parallel gradient descent algorithm [33]. It is noteworthy that the performance of the FSO transmission system is dominated by linear mode crosstalk. within comparison to fiber-optic transmission systems, FSO links will not have notable nonlinear effects in principle. The AO used in our experiment will not bring nonlinear distortions, since the processing time of the AO is far longer than the data rate.

Following AO compensation, the less distorted OAM modes are detected by SLM4. This is used to convert one OAM mode back to the Gaussian-like mode, which is then collected by another SMF patch cable. The collected optical signals are pre-amplified by EDFA2, followed by a variable optical attenuator (VOA) for optical power tuning. Additional ASE noise is generated and adjusted via a sub-system configured by EDFA3, EDFA4, and VOA2. Such ASE noise is added to the optical signal using an optical coupler, after which an OTF with the central wavelength of 1550.12 nm is applied to single out the corresponding optical wavelength channel. In the coherent receiver, the local oscillator (LO) light and the optical signal are mixed in an optical 90° hybrid, detected by two PDs, and digitized by a real-time oscilloscope. After the off-line digital signal processing (DSP) signal recovery, the sum-product algorithm is used in the LDPC decoding procedure with a maximum of 50 iterations [34].

2.2. Results and Analysis

We begin by investigating the atmospheric turbulence-induced mode crosstalk and the merits of AO assisted wavefront correction. The power ratio between the target OAM modes (OAM states ±2, ±6) and their adjacent modes are used here as a metric to evaluate the effect of mode crosstalk. As illustrated in Figure 2a, the power of target OAM mode measured was similar to the adjacent OAM modes, indicating that the data originating from mode crosstalk will severely interfere with the desirable data after mode detection. Data in Figure 2b shows that the average extinction ratio (ER) after AO assisted wavefront correction reaches a 6-dB improvement. The blue bars in Figure 2a and navy bars in Figure 2b represent the desirable transmitted OAM modes, and the green bars represents the unwanted OAM modes caused by OAM mode crosstalk.

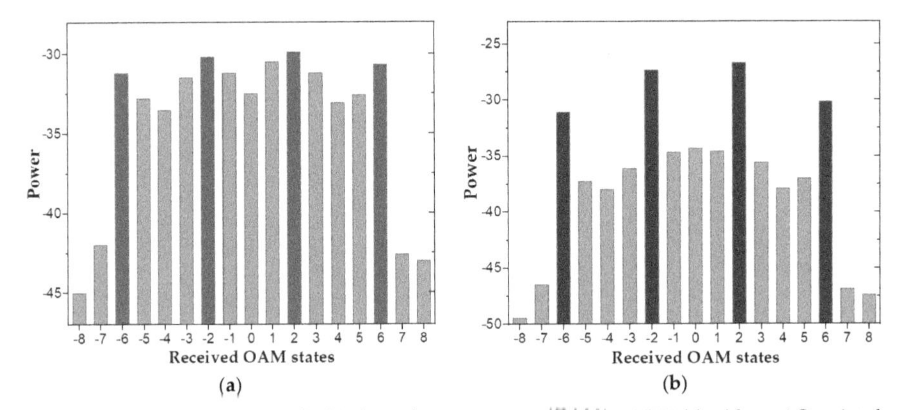

Figure 2. Power distributions of orbital angular momentum (OAM) modes: (**a**) without AO assisted wavefront correction and (**b**) with AO assisted wavefront correction.

Figure 3a shows the average bit-error rate (BER) vs. OSNR performance with or without atmospheric turbulence effects. The data clearly demonstrates that the BER curves do not drop quickly, even with the increasing OSNR values. It is caused by the unperfect mode generation and detection patterns, which will also introduce unwanted inter-mode crosstalk effects. Note that the worse BER performance of OAM states ±6 compared to OAM state ±2 is due to high-order OAM mode sensitivity to the boundary effect.

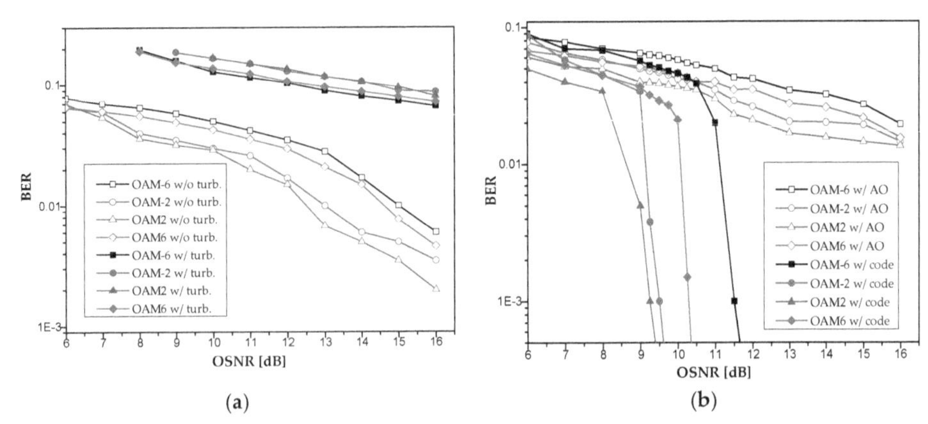

Figure 3. (**a**) Average BER vs. OSNR performance in cases with or without atmospheric turbulence effects. (**b**) Average BER vs. OSNR performance after AO assisted wavefront correction and low-density parity-check (LDPC) coding.

Figure 3b shows that distinct average OSNR gain can be reached after AO assisted wavefront correction and LDPC coding. We observe that the BER curves after AO assisted wavefront correction can be lower than 0.04, which is the error correction threshold of the used LDPC code. Furthermore, the performance of the AO assisted wavefront correction on OAM states ± 2 are better than that of OAM states ± 6. This is caused not only by the boundary effect brought by the limited sizes of the used optical components, but also the features of the used probe beam. This beam is able to fully cover the small-size OAM modes (OAM states ± 2) after the turbulent FSO transmission, rather than the large-size OAM modes (OAM states ± 6). After the AO correction is performed, LDPC coding/decoding can then be applied to efficiently eliminate the post-FEC error floor. Figure 3b shows that the BER curves of all OAM modes can drop quickly as long as the OSNR is higher than 8 dB. More specifically, when the BER is 2×10^{-2} the coding gains of 3.9, 4.1, 5.2, and 5 dB are reached for OAM states 2, -2, 6, and -6, respectively.

In this study, we implement pre-compensation algorithms at the transmitter side and post-equalization algorithms at the receiver side to minimize the implementation penalty. Turning off the turbulence emulator by sending blank phase patterns to the SLMs in the turbulence emulator does not eliminate all OAM mode distortions caused by the imperfect SLM screens, especially at the OAM mode generation and detection steps.

3. Joint Huffman and LDPC Coding Enabled Free-Space Optical Communication

3.1. Experimental Setup

Figure 4 shows the experimental setup for joint Huffman and LDPC coding enabled FSO communication system. In the transmitter, a 1550 nm CW light is generated as the optical carrier, and passes to an optical I/Q modulator. The PRBS signals are coded by Huffman procedure to the symbol sequences with alphabet sizes of 5 and 9 respectively [30], or uniformly mapped to symbol sequences with alphabet sizes of 4 and 8, respectively. Then the GF(5), GF(3^2) based nonbinary LDPC encoding is used, followed by mapping procedures from the coded sequences to 5-QAM and 9-QAM signals, respectively. Classical binary LDPC encoding is implemented for QPSK (or 4-QAM) and 8-QAM sequences. The used Huffman trees and the 5/9-QAM formats with corresponding bit labeling are provided (Figure 4(a1,a2,b1,b2)). When 12.5 G Baud electronic signals are generated, they drive the I/Q modulator. The optical signals are boosted by EDFA1, and filtered with an OTF. Then optical signals are separated by an optical coupler, decorrelated, and converted to OAM states 2 and -6 by SLM1. The formed OAM modes are combined and centrally aligned by BS1. BS2 is used to separate the multiplexed OAM modes, and re-combined by the BS3. The optical signals in one optical path bounce off a mirror one time to generate the opposite OAM modes, i.e., OAM states -2 and 6. The optical signals in the other path bounce off four times to keep the original mode states and decorrelate the carried optical signals. The desired SO between the two optical paths can be achieved by adjusting the BS3 position. In this setup, the limited SLM screen size restricts the SO freedom. To reduce the side effect-induced inter-mode crosstalk and mode power loss, OAM states 2 and -6 are centrally aligned and transmitted in one optical path, while OAM states -2 and 6 are combined and launched onto another optical path.

The offset OAM beams are expanded by a BE, and then distorted in the designed turbulence emulator. The current turbulence emulator is designed according to the Rytov variance of 0.5. The distorted OAM beams are captured by a compressing telescope, demultiplexed by SLM4, and back-convert the target OAM mode to the Gaussian mode. The Gaussian beam is then coupled from free space into a fiber cable and pre-amplified by EDFA2. The ASE noise is loaded onto the signal in the 3-dB coupler, and later the out-of-band noise is removed by the OTF2. The optical signal is detected by a coherent receiver, and equalized by DSP signal recovery. The symbol log-likelihood ratio (LLR) estimation is executed before the GF(2)/GF(5)/GF(3^2) based LDPC decoding procedures. Ultimately, BER values are calculated after LDPC decoding to determine system performance.

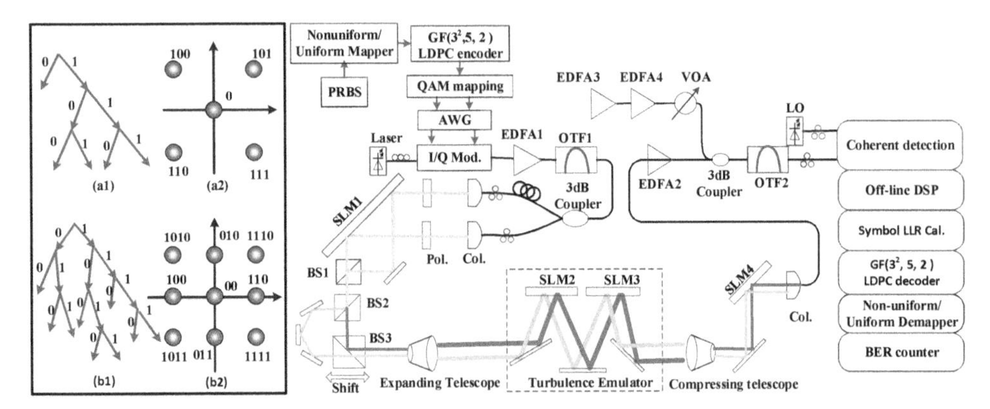

Figure 4. Experimental setup for the joint Huffman and LDPC coding enabled FSO communication system. Insets: (**a1**) Huffman tree for 5-size alphabet; (**a2**) 5-QAM format with bit labeling; (**b1**) Huffman tree for 9-size alphabet; (**b2**) 9-QAM format with bit labeling.

3.2. Results and Analysis

The signal-to-crosstalk ratio (SCTR) gains achieved by SO are presented in Figure 5. The SCTR is measured after the OAM mode detection. It is defined as the power ratio of the desirable OAM mode and neighboring OAM modes, in the scenario where the desired OAM mode is generated exclusively at the transmitter. The insets in Figure 5(a1,a2) show the measured SCTR gains. The SCTR improvement of >1.6 dB is achieved for OAM state 2; and the SCTR gain of >1 dB is available for OAM state −6. The data shows that the best SO is at 6 mm and 5 mm for OAM states 2, and −6, respectively. For simplicity, the SO will be set to 5 mm for all following cases in this paper.

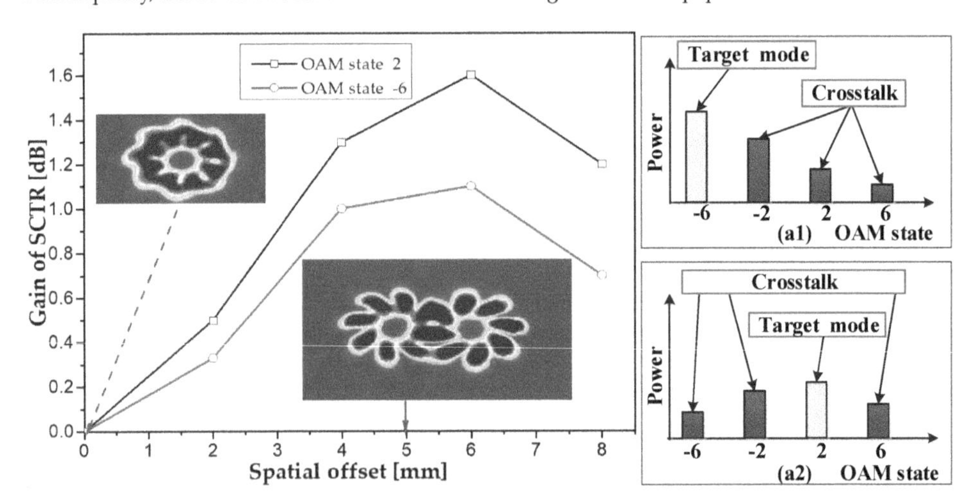

Figure 5. Signal-to-crosstalk ratio (SCTR) improvements obtained by the spatial offset (SO). Insets: The examples of SCTR calculation for (**a1**) OAM state −6, and (**a2**) OAM state 2.

In Figure 6, we analyze the effects of the nonuniform signaling. The data represents BER performance in the atmospheric turbulence-free environment. Figure 6a,b shows that the pre-FEC performance of 5-QAM is worse than that of the QPSK due to the implementation and DSP penalties. These data also indicate that the pre-FEC performance of 9-QAM can outperform the 8-QAM. However, the post-FEC OSNR penalties between the 5-QAM and QPSK are measured to be <0.3 dB and 0.2 dB, in the respective OAM states 2 and −6, when the BER is 10^{-4}. In addition, GF(3^2) LDPC encoded

9-QAM shows a better performance over GF(2) LDPC coded 8-QAM by 2.7 dB and 3.2 dB in respective OAM states 2 and −6, when BER is 10^{-4}.

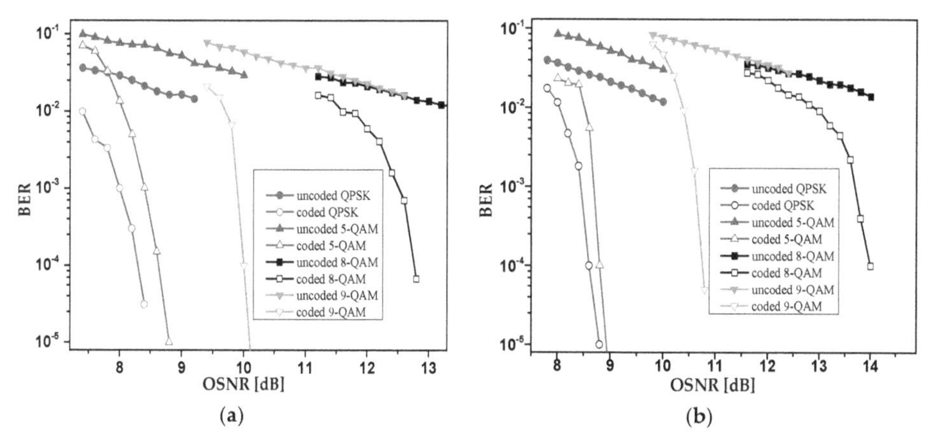

Figure 6. BER vs. OSNR performance in the atmospheric turbulence-free environment, when (**a**) OAM state 2 is under testing; (**b**) OAM state −6 is under testing.

The BER performance affected by atmospheric turbulence is shown in Figure 7. The data indicate that the pre-FEC BER performance gap between QPSK and 5-QAM shrinks in an atmospheric turbulence limited environment; while the pre-FEC BER performance of 9-QAM is better than that that of the 8-QAM. Figure 7a shows that, if OAM state 2 is under testing, the average OSNR gains of >1.6 dB and 5.6 dB are reached by 5-QAM and 9-QAM, respectively, when the post-FEC BER is 10^{-4}, compare to coded QPSK and 8-QAM. In addition, if OAM state −6 is detected, as depicted in Figure 7b, the coding gains of >1.1 dB and 5.4 dB can be obtained when the post-FEC BER is 10^{-4} respectively by comparing 5-QAM with QPSK, and 9-QAM with 8-QAM. It is noteworthy that we only measure the performances of OAM states 2 and −6, this is due to the symmetry between OAM modes with positive and negative states. In other words, OAM state 2 will have a very similar performance as OAM state −2; while OAM state 6 will also have a similar performance as OAM state −6.

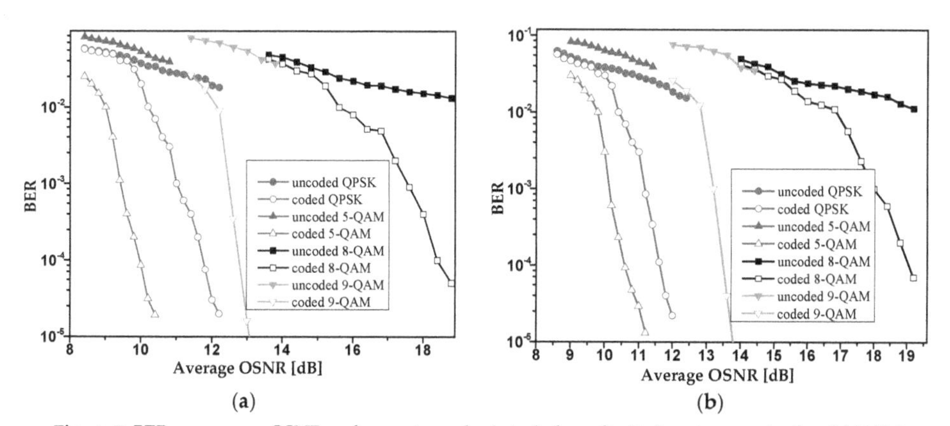

Figure 7. BER vs. average OSNR under an atmospheric turbulence limited environment, when (**a**) OAM state 2 is under testing; (**b**) OAM state −6 is under testing.

In order to clarify the performance improvements achieved by the coded modulation, the average OSNR requirements at BER = 10^{-4} for 4/5/8/9-QAM formats are shown in Table 1. These data are under conditions where OAM states 2 and 6 are detected, and with/without atmospheric turbulence. The OSNR penalties brought by atmospheric turbulence for OAM state 2, are concluded as 3.3 dB, 1.4 dB, 5.7 dB, and 2.8 dB for 4/5/8/9-QAM, respectively. Similarly, the mode-crosstalk penalties for OAM state −6, are measured to be 3.1 dB, 1.8 dB, 5 dB, and 2.8 dB for 4/5/8/9-QAM, respectively. Thereafter, the nonuniform 5/9-QAM schemes have a higher inter-mode crosstalk tolerance over uniform QPSK and 8-QAM.

Table 1. Minimum OSNR requirements at BER of 10^{-4}.

	Modulation Formats	QPSK	5-QAM	8-QAM	9-QAM
Without turbulence	OAM state 2	8.3 dB	8.6 dB	12.7 dB	10 dB
	OAM state 6	8.6 dB	8.8 dB	14.0 dB	10.8 dB
With turbulence	OAM state 2	11.6 dB	10 dB	18.4 dB	12.8 dB
	OAM state 6	11.7 dB	10.6 dB	19.0 dB	13.6 dB

4. Concluding Remarks

We have investigated the high-speed OAM multiplexed FSO communication system, enabled by AO based wavefront correction and LDPC coding. The inter-mode crosstalk was first compensated by the wavefront sensorless AO setup, and the residual mode crosstalk induced data interference was later solved by sufficiently strong LDPC coding.

We also presented a crosstalk-resistance solution in an OAM multiplexed FSO link based on SO and coded modulation. More than 1 dB SCTR improvement has been shown for the used OAM modes. Moreover, the 5/9-QAM schemes exhibit a better crosstalk tolerance than regular QPSK and 8QAM schemes.

Sometimes, more advanced modulation formats are used to further increase channel capacity. When high-order modulation formats are used, FSO communication systems are more sensitive to atmospheric turbulence. Some turbulence compensation solutions may not work well. Stronger FEC coding may be used to protect systems reliability. Optical domain turbulence compensation solutions, like adaptive optics, are suggested for implementation before mode detection. Nonuniform coded modulation may still work, but should not bring much performance improvement like 5/9QAM formats. MIMO processing solutions will face a more severe challenge, due to the increasing computation complexity and the reduced robustness of the MIMO equalization.

Author Contributions: This research was supervised by I.B.D. All laboratory work was done by Z.Q.

Funding: This research was funded by ONR grant number [N00014-13-1-0627].

Conflicts of Interest: The authors declare no conflicts of interest.

References

1. Qu, Z.; Li, Y.; Mo, W.; Yang, M.; Zhu, S.; Kilper, D.; Djordjevic, I.B. Performance optimization of PM-16QAM transmission system enabled by real-time self-adaptive coding. *Opt. Lett.* **2017**, *42*, 4211–4214. [CrossRef] [PubMed]
2. Ip, E.; Lau, A.P.T.; Barros, D.J.F.; Kahn, J.M. Coherent detection in optical fiber systems. *Opt. Express* **2008**, *16*, 753–791. [CrossRef] [PubMed]
3. Khalighi, M.A.; Uysal, M. Survey on free space optical communication: A communication theory perspective. *IEEE Commun. Surv. Tutor.* **2014**, *16*, 2231–2258. [CrossRef]
4. Qu, Z.; Djordjevic, I.B. Experimental evaluation of LDPC-coded OAM based FSO communication in the presence of atmospheric turbulence. In Proceedings of the 12th International Conference on Telecommunications in Modern Satellite, Cable and Broadcasting Services (TELSIKS), Nis, Serbia, 14–17 October 2015; pp. 117–122.

5. Djordjevic, I.B.; Qu, Z. Coded Orbital Angular Momentum Modulation and Multiplexing Enabling Ultra-High-Speed Free-Space Optical Transmission. In *Optical Wireless Communications Signals and Communication Technology*; Uysal, M., Capsoni, C., Ghassemlooy, Z., Boucouvalas, A., Udvary, E., Eds.; Springer: Cham, Switzerland, 2016; pp. 363–385. ISBN 978-3-319-30200-3.
6. Yue, Y.; Huang, H.; Ahmed, N.; Yan, Y.; Ren, Y.; Xie, G.; Rogawski, D.; Tur, M.; Willner, A.E. Reconfigurable switching of orbital-angular-momentum-based free-space data channels. *Opt. Lett.* **2013**, *38*, 5118–5121. [CrossRef] [PubMed]
7. Jurado-Navas, A.; Tatarczak, A.; Lu, X.; Olmos, J.J.; Garrido-Balsells, J.M.; Monroy, I.T. 850-nm hybrid fiber/free-space optical communications using orbital angular momentum modes. *Opt. Express* **2015**, *23*, 33721–33732. [CrossRef] [PubMed]
8. Xie, Y.; Geng, Z.; Kong, D.; Zhuang, L.; Lowery, A.J. Selectable-FSR 10-GHz granularity WDM superchannel filter in a reconfigurable photonic integrated circuit. *J. Lightw. Technol.* **2018**, *36*, 2619–2626. [CrossRef]
9. Xie, Y.; Geng, Z.; Zhuang, L.; Burla, M.; Taddei, C.; Hoekman, M.; Leinse, A.; Roeloffzen, C.G.; Boller, K.J.; Lowery, A.J. Programmable optical processor chips: Toward photonic RF filters with DSP-level flexibility and MHz-band selectivity. *Nanophotonics* **2017**, *7*, 421–454. [CrossRef]
10. Lowery, A.J.; Xie, Y.; Zhu, C. Systems performance comparison of three all-optical generation schemes for quasi-Nyquist WDM. *Opt. Express* **2015**, *23*, 21706–21718. [CrossRef] [PubMed]
11. Qu, Z.; Zhang, S.; Djordjevic, I.B. Universal Hybrid Probabilistic-geometric Shaping Based on Two-dimensional Distribution Matchers. In Proceedings of the Optical Fiber Communication Conference (OFC), San Diego, CA, USA, 11–15 March 2018.
12. Zhang, S.; Qu, Z.; Yaman, F.; Mateo, E.; Inoue, T.; Nakamura, K.; Inada, Y.; Djordjevic, I.B. Flex-Rate Transmission using Hybrid Probabilistic and Geometric Shaped 32QAM. In Proceedings of the Optical Fiber Communication Conference (OFC), San Diego, CA, USA, 11–15 March 2018.
13. Lin, C.; Qu, Z.; Liu, T.; Zou, D.; Djordjevic, I.B. Experimental study of capacity approaching general LDPC coded non uniform shaping modulation format. In Proceedings of the Asia Communications and Photonics Conference (ACP), Wuhan, China, 2–5 November 2016.
14. Wang, J.; Yang, J.Y.; Fazal, I.M.; Ahmed, N.; Yan, Y.; Huang, H.; Ren, Y.; Yue, Y.; Dolinar, S.; Tur, M.; et al. Terabit free-space data transmission employing orbital angular momentum multiplexing. *Nat. Photonics* **2012**, *6*, 488–496. [CrossRef]
15. Djordjevic, I.B.; Qu, Z. *Coded Orbital-Angular-Momentum-Based Free-Space Optical Transmission*; Wiley Encyclopedia of Electrical and Electronics Engineering: Hoboken, NJ, USA, 2016.
16. Willner, A.E.; Huang, H.; Yan, Y.; Ren, Y.; Ahmed, N.; Xie, G.; Bao, C.; Li, L.; Cao, Y.; Zhao, Z.; et al. Optical communications using orbital angular momentum beams. *Adv. Opt. Photonics* **2015**, *7*, 66–106. [CrossRef]
17. Allen, L.; Beijersbergen, M.W.; Spreeuw, R.J.C.; Woerdman, J.P. Orbital angular momentum of light and the transformation of Laguerre-Gaussian laser modes. *Phys. Rev. A* **1992**, *45*, 8185. [CrossRef] [PubMed]
18. Zhang, Y.; Shan, L.; Li, Y.; Yu, L. Effects of moderate to strong turbulence on the Hankel-Bessel-Gaussian pulse beam with orbital angular momentum in the marine-atmosphere. *Opt. Express* **2017**, *25*, 33469–33479. [CrossRef]
19. Qu, Z.; Djordjevic, I.B. Beyond 1 Tb/s free-space optical transmission in the presence of atmospheric turbulence. In Proceedings of the Photonics North (PN), Ottawa, ON, Canada, 6–8 June 2017.
20. Ren, Y.; Huang, H.; Xie, G.; Ahmed, N.; Yan, Y.; Erkmen, B.I.; Chandrasekaran, N.; Lavery, M.P.; Steinhoff, N.K.; Tur, M.; et al. Atmospheric turbulence effects on the performance of a free space optical link employing orbital angular momentum multiplexing. *Opt. Lett.* **2013**, *38*, 4062–4065. [CrossRef] [PubMed]
21. Ren, Y.; Xie, G.; Huang, H.; Li, L.; Ahmed, N.; Yan, Y.; Lavery, M.P.; Bock, R.; Tur, M.; Neifeld, M.A.; et al. Turbulence compensation of an orbital angular momentum and polarization-multiplexed link using a data-carrying beacon on a separate wavelength. *Opt. Lett.* **2015**, *40*, 2249–2252. [CrossRef] [PubMed]
22. Qu, Z.; Djordjevic, I.B. 500 Gb/s free-space optical transmission over strong atmospheric turbulence channels. *Opt. Lett.* **2016**, *41*, 3285–3288. [CrossRef] [PubMed]
23. Li, S.; Wang, J. Adaptive free-space optical communications through turbulence using self-healing Bessel beams. *Sci. Rep.* **2017**, *7*, 43233. [CrossRef] [PubMed]

24. Huang, H.; Cao, Y.; Xie, G.; Ren, Y.; Yan, Y.; Bao, C.; Ahmed, N.; Neifeld, M.A.; Dolinar, S.J.; Willner, A.E. Crosstalk mitigation in a free-space orbital angular momentum multiplexed communication link using 4× 4 MIMO equalization. *Opt. Lett.* **2014**, *39*, 4360–4363. [CrossRef] [PubMed]
25. Ren, Y.; Wang, Z.; Xie, G.; Li, L.; Cao, Y.; Liu, C.; Liao, P.; Yan, Y.; Ahmed, N.; Zhao, Z.; et al. Free-space optical communications using orbital-angular-momentum multiplexing combined with MIMO-based spatial multiplexing. *Opt. Lett.* **2015**, *40*, 4210–4213. [CrossRef] [PubMed]
26. Dabiri, M.T.; Saber, M.J.; Sadough, S.M.S. On the performance of multiplexing FSO MIMO links in log-normal fading with pointing errors. *J. Opt. Commun. Netw.* **2017**, *9*, 974–983. [CrossRef]
27. Qu, Z.; Djordjevic, I.B. Coded orbital angular momentum based free-space optical transmission in the presence of atmospheric turbulence. In Proceedings of the Asia Communications and Photonics Conference (ACP), Hong Kong, China, 19–23 November 2015.
28. Kaushal, H.; Kaddoum, G. Optical communication in space: Challenges and mitigation techniques. *IEEE Commun. Surv. Tutor.* **2017**, *19*, 57–96. [CrossRef]
29. Qu, Z.; Djordjevic, I.B. Approaching terabit optical transmission over strong atmospheric turbulence channels. In Proceedings of the 18th International Conference on Transparent Optical Networks (ICTON), Trento, Italy, 10–14 July 2016; pp. 1–5.
30. Liu, T.; Qu, Z.; Lin, C.; Djordjevic, I.B. Nonuniform signaling based LDPC-coded modulation for high-speed optical transport networks. In Proceedings of the Asia Communications and Photonics Conference (ACP), Wuhan, China, 2–5 November 2016.
31. Qu, Z.; Lin, C.; Liu, T.; Djordjevic, I.B. Experimental study of nonlinearity tolerant modulation formats based on LDPC coded non-uniform signaling. In Proceedings of the Optical Fiber Communication Conference (OFC), Los Angeles, CA, USA, 19–23 March 2017.
32. Andrews, L.C.; Phillips, R.L.; Hopen, C.Y. *Laser Beam Scintillation with Applications*; SPIE Press: Bellingham, WA, USA, 2001; ISBN 9781510604896.
33. Dong, B.; Ren, D.Q.; Zhang, X. Stochastic parallel gradient descent based adaptive optics used for a high contrast imaging coronagraph. *Res. Astron. Astrophys.* **2011**, *11*, 997. [CrossRef]
34. Hu, X.Y.; Eleftheriou, E.; Arnold, D.M., Dholakia, A. Efficient implementations of the sum-product algorithm for decoding LDPC codes. In Proceedings of the Global Telecommunications Conference, San Antonio, TX, USA, 25–29 November 2001.

Article

An Evaluation of Orbital Angular Momentum Multiplexing Technology

Doohwan Lee *, Hirofumi Sasaki, Hiroyuki Fukumoto, Yasunori Yagi and Takashi Shimizu

NTT Network Innovation Laboratories, NTT Corporation, 1-1 Hikarinooka, Yokosuka-Shi 239-0847, Japan; hirofumi.sasaki.uw@hco.ntt.co.jp (H.S.); hiroyuki.fukumoto.mp@hco.ntt.co.jp (H.F.); yasunori.yagi.zc@hco.ntt.co.jp (Y.Y.); takashi.shimizu.nu@hco.ntt.co.jp (T.S.)

* Correspondence: doohwan.lee.yr@hco.ntt.co.jp; Tel.: +81-46-859-4778

Received: 8 March 2019; Accepted: 24 April 2019; Published: 26 April 2019

Abstract: This paper reports our investigation of wireless communication performance obtained using orbital angular momentum (OAM) multiplexing, from theoretical evaluation to experimental study. First, we show how we performed a basic theoretical study on wireless OAM multiplexing performance regarding modulation, demodulation, multiplexing, and demultiplexing. This provided a clear picture of the effects of mode attenuation and gave us insight into the potential and limitations of OAM wireless communications. Then, we expanded our study to experimental evaluation of a dielectric lens and end-to-end wireless transmission on 28 gigahertz frequency bands. To overcome the beam divergence of OAM multiplexing, we propose a combination of multi-input multi-output (MIMO) and OAM technology, named OAM-MIMO multiplexing. We achieved 45 Gbps (gigabits per second) throughput using OAM multiplexing with five OAM modes. We also experimentally demonstrated the effectiveness of the proposed OAM-MIMO multiplexing using a total of 11 OAM modes. Experimental OAM-MIMO multiplexing results reached a new milestone for point-to-point transmission rates when 100 Gbps was achieved at a 10-m transmission distance.

Keywords: orbital angular momentum multiplexing; OAM; OAM-MIMO; 28 GHz; uniform circular array; dielectric lens

1. Introduction

Recently, wireless communication using OAM (Orbital Angular Momentum) has drawn much attention as an emerging candidate for beyond 5G (fifth generation) technology due to its potential as a means to enable high-speed wireless transmission. OAM is a physical property of electro-magnetic waves that are characterized by a helical phase front in the propagation direction. Since the characteristic can be used to create multiple independent channels, wireless OAM multiplexing can effectively increase the transmission rate in a point-to-point link such as wireless backhaul and/or fronthaul [1,2]. Recent seminar work demonstrated the feasibility of OAM multiplexing by achieving 32 Gbps (gigabits per second) transmission, as Yan et al. reported using the 28 GHz (gigahertz) band in 2014 [1] and the 60 GHz band in 2016 [3]. Since OAM multiplexing technology is relatively new, it is important to validate the feasibility from various perspectives. To do that, we first validated the feasibility from a theoretical perspective using simulations (Section 2). We then validated the feasibility from beam generation and propagation perspectives in experiments (Section 3). Finally, we concluded by validating the feasibility from the end-to-end wireless communication perspective using experiments (Section 4). In our previous research, we explored the potential of wireless OAM multiplexing by conducting the following three studies.

The first part of our work was theoretically investigating the feasibility of OAM multiplexing. First, we investigated the theoretical performance of modulation, demodulation, multiplexing, and demultiplexing OAM algorithms using computer simulations [4]. This enabled us to clarify the performance and effect of mode-dependent power attenuation. In doing so we generated OAM signals by using a UCA (uniform circular array) that comprises multiple omnidirectional antenna elements.

The second part of our work was validating the feasibility from beam generation and propagation perspectives. In particular, we expanded our study to the usage of a dielectric lens to examine the feasibility of long distance transmission using OAM [5]. Ideally, all OAM modes are orthogonal to each other due to the unique nature of their phase fronts. However, with these modes, it is difficult to transmit over long distances because their energy rapidly diverges as the beam propagates. To achieve long-distance transmission, we developed and proposed a beam divergence reduction method using the focusing effect of a dielectric lens. We conducted a wave propagation experiment on 28 GHz bands to demonstrate the effectiveness of using such a lens. In the experiment, we were able to generate OAM modes 0, ±1, and ±2. In addition, we showed the beam divergence reduction effect by making a comparison between conventional OAM beam generation methods and using a dielectric lens with a UCA.

In the third part of our work, we validated the feasibility from the end-to-end wireless communication perspective. We report experimental results using wireless OAM multiplexing at 28 GHz. One of the major challenges for achieving OAM multiplexing is its intensity variation among different OAM modes. As shown in Figure 7, intensity distributions of OAM signals with different modes are given by the Bessel function of the first kind by their inherent nature [2]. This yields selective reception (Rx) SNR (signal-to-noise ratio) degradation when the Rx antenna is located in a low SNR region (null region). To address this problem, we used multiple UCAs, which are designed to avoid reception in null regions. We obtained successful experimental results using five OAM modes over 28 GHz [6]. We also developed and proposed OAM-MIMO multiplexing using multiple UCAs. Unlike OAM multiplexing, OAM-MIMO multiplexing exploits multiple sets of the same OAM modes with receiver equalizations [7]. This enables the number of concurrently transmitted data streams to be increased without using higher OAM modes that have large beam divergence. We experimentally demonstrated the effectiveness of the proposed OAM-MIMO multiplexing by using 11 OAM modes in total (three OAM modes 0 and two sets each of OAM modes ±1 and ±2). Experimental results reached a new milestone in point-to-point transmission rates when 100 Gbps was achieved at 10 m transmission distance.

2. Background and Theoretical Performance Evaluation

2.1. OAM Generation Using a Uniform Circular Array

Studies regarding OAM multiplexing in the wireless communication field are categorized into antenna design and beam generation, end-to-end experiments, signal processing methods, and system studies for topics such as capacity analysis and link budget. Among these, we mainly focus on antenna design in this subsection. Various types of antenna designs have been reported using helicoidally deformed parabolic antennas [8], spiral phase plates (SPP) [2,3,9], holographic plates [10], elaborately tuned planer SPPs [11], and other components [12,13]. In our latest research, we focused on OAM generation using UCAs [14–16]. Figure 1 shows OAM mode generation by using UCAs. The phase of each antenna is shifted in accordance with an OAM mode. The transmitted signal from each antenna can be written in vector form as

$$x = \left[1, e^{j\frac{2\pi L}{N}}, \cdots, e^{j\frac{2\pi(N-1)}{N}}\right]^T, \tag{1}$$

where L is the OAM mode number and N is the number of radiating antennas in the transmitting UCA. By using a UCA, the OAM state number L is limited by the number of transmitting antennas as $|L| < N/2$.

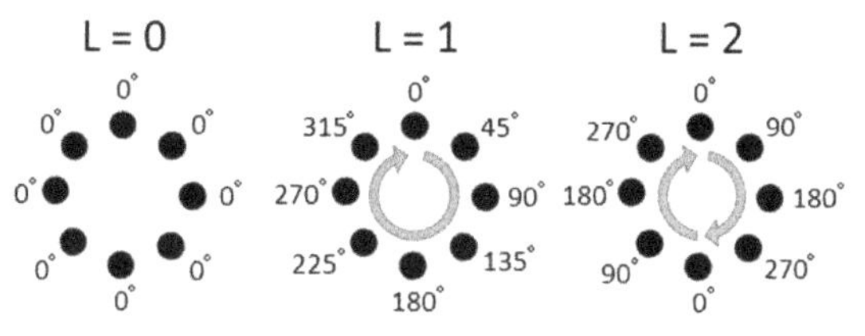

Figure 1. Generation of OAM modes by a uniform circular array.

Since the diffraction pattern of the UCA can be approximated by the Bessel beam [17], the field distribution of the beam with the OAM mode L is often expressed in the Bessel beam equation as

$$v_L(r,\theta,z) = \frac{\lambda \exp[(2\pi i/\lambda)\sqrt{r^2+z^2}]}{4\pi\sqrt{r^2+z^2}} \cdot i^{-L}\exp[iL\theta] \cdot J_L\left(\frac{2\pi rD}{\lambda\sqrt{r^2+z^2}}\right), \tag{2}$$

where $J_L(\cdot)$, λ, and D, respectively, denote the Lth order Bessel function of the first kind, the wavelength of the carrier frequency, and the radius of the transmitting UCA. Equation (2) is represented in cylindrical coordinates, where r and θ are, respectively, radius and azimuthal angle at the Rx plane that is vertical to the beam propagation direction. z is the distance between the centers of the Tx (transmission) and Rx UCAs. Here, let us discuss a comparison between OAM and MIMO. OAM is a specific implementation of MIMO with circular antenna arrays. The difference is that an OAM beam can be achieved virtually and practically in accordance with an OAM state, so it is possible to design a mechanism to manipulate that beam specifically to achieve a specific type of communication. The difference from the conventional MIMO and OAM multiplexing is as follow. To obtain a full rank matrix, usually an NLOS (non-line-of-sight) multipath and a rich scattering environment are assumed in the conventional MIMO. In LOS (line-of-sight) environment cases, digital signal processing such as precoding at the transmitter may be required for the conventional MIMO case to obtain a full rank matrix. On the other hand, OAM does not need digital signal processing at the transmitter to obtain the full rank matrix.

2.2. Modulation and Demodulation

Modulation and demodulation using OAM can be mainly categorized into two schemes. We detail these schemes as follows.

- OAM Shift Keying (OAMSK) [14]: This scheme simply puts binary data into an OAM mode. For example, bit "0" is mapped as OAM mode 1, while bit "1" is mapped as mode -1 (minus 1). OAMSK modulated signals can be demodulated by using the phase gradient method, an FFT (fast Fourier transform) based method, or ML (maximum likelihood) detection. The gradient method uses the phase difference between two receiving antennas to determine the OAM mode. The FFT-based method conducts the FFT process using a reception (Rx) UCA and chooses the maximum coefficients. ML detection selects the OAM mode with the closest distance to the received signal.
- OAM Division Multiplexing (OAMDM) [16]: This scheme uses OAM modes to carry multiple streams of data simultaneously. An OAM mode can carry one stream, similar to the way that one OFDM (orthogonal frequency division multiplexing) subcarrier can. This scheme potentially improves the spectrum efficiency. With it, OAMDM modulated signals are demodulated similar to the way they are with MIMO equalization techniques such as zero forcing or minimum mean square error equalization, assuming the channel information is available. Since OAM multiplexing is expected to be used under LOS environments with static channels such as wireless fronthaul/backhaul, simplified channel estimation using Equation (2) might be feasible.

2.3. Mode-Dependent Power Distribution

In the work we report here, we also considered two key issues regarding the mode-dependent power distribution among different OAM modes. These issues are as follows.

- Peak Rx Power Degradation: As the number of OAM modes increases, the radiation becomes wider, the angle from the beam axis at the peak Rx power becomes wider, and the SNR at its peak Rx power becomes smaller. Accordingly, the performance is degraded as the number of OAM modes increases.
- Non-identical Peak Rx Power Locations: The peak Rx power locations of each OAM mode are not identical because their radiation patterns are distinct. Therefore, the mode-dependent performance degradation becomes more severe when a single Rx UCA is used because some OAM modes might not have the peak Rx power at a certain location.

2.4. Evaluation

We implemented a simulation testbed of OAM based wireless communication at 60 GHz. Figure 1 shows an illustration of the generation of OAM signals using UCA. The gain for each antenna element reflects the UCA as set to be 0 dBi (decibels relative to isotropic radiator). Note that concurrent transmission of multiple OAM modes can be achieved by superposing multiple OAM signals. It is generally assumed that OAM multiplexing is to be mainly used in LOS environments such as wireless fronthaul and/or backhaul. Therefore, we used an AWGN (additive white Gaussian noise) channel environment.

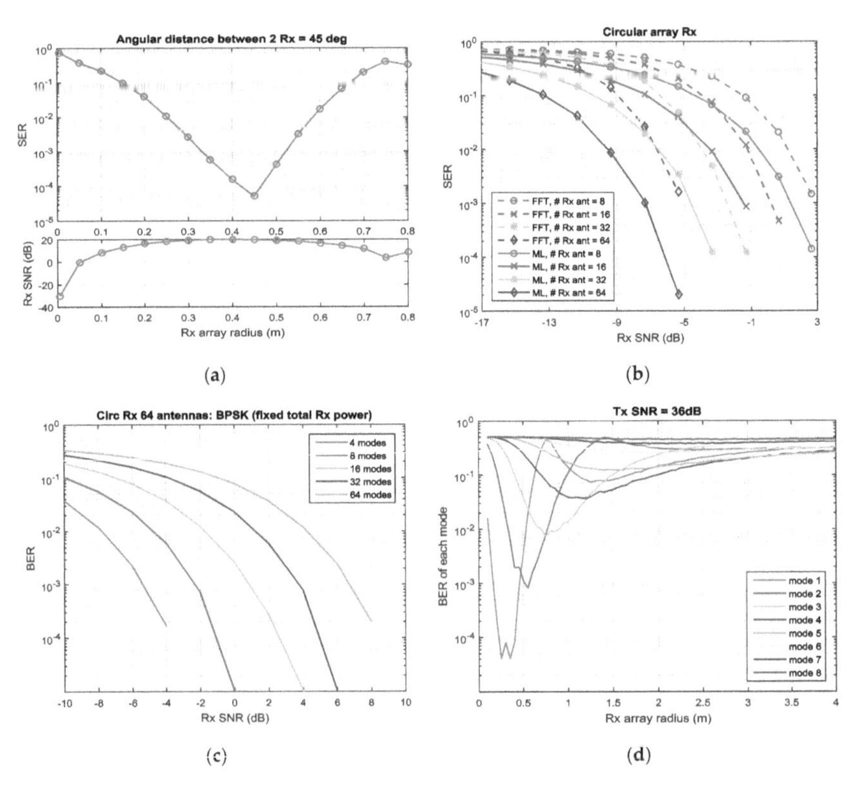

Figure 2. Performance evaluations of: (**a**) OAMSK (phased gradient method with varying Rx array radius); (**b**) OAMSK (FFT-based and ML methods); (**c**) OAMDM with fixed Rx power; and (**d**) OAMDM with varying Rx array radius.

Figure 2a shows the OAMSK performance obtained by the phase gradient method, which uses two antennas for OAM signal detection with varying Rx array radius and angular distance between two Rx antenna elements. Although the performance is poorer than that obtained with FFT-based and ML methods, only two antenna elements are necessary. This is in contrast to cases in which all the Rx UCA elements are required. This is favorable for higher OAM mode transmission. Figure 2b shows the OAMSK performance obtained by using the FFT-based method and ML detection while varying the number of antenna components in the Rx UCA. We found that in both cases the ML detection yields generally better results and additional performance gain is achieved as the number of antenna components in the Rx UCA increases. Figure 2c shows the OAMDM performance with fixed total Rx power among OAM modes. Note that Rx signals in this curve are obtained at the location of the peak Rx power of each mode. In this case, we observed performance degradation of 3 dB as the number of OAM modes increased. Figure 2d shows the effect of a non-identical location of the peak Rx power by varying the Rx array radius. As the OAM mode number increases, the Rx array radius for the best BER (bit error rate) performance also increases. Correspondingly, if the Rx array radius is customized for a certain OAM mode, the performance of other OAM modes' signals might be deteriorated severely. For this, multiple UCAs might be necessary, which leads us to use multiple UCAs.

In this subsection, we report how we studied the potential and limitations of OAM-based wireless communication in terms of mode-dependent power attenuation and non-identical peak Rx power locations through the use of modulation and demodulation algorithms. We confirmed the effect of mode-dependent performance variations. Further study is necessary to fully rectify the undesirable effects of the mode-dependent power attenuation.

3. Beam Focusing Effect Using Dielectric Lens for OAM Multiplexing

3.1. Usage of Dielectric Lens for OAM Multiplexing

We present the beam focusing effect obtained by using a dielectric lens for OAM multiplexing to increase the transmission distance. Figure 3 shows the configuration of the transmitting and receiving antennas. The OAM mode radiates from the UCA installed on the back of the dielectric lens. The OAM mode radiated into the space is phase-modulated by the lens and reaches the reception point. When transmitting the OAM mode n, we weight each element of the array antenna. Since the phase distribution in the OAM mode is circularly symmetric in the plane perpendicular to the beam traveling direction, a circular lens is used so that circular symmetry is not disturbed by phase modulation at the time of passing through the lens.

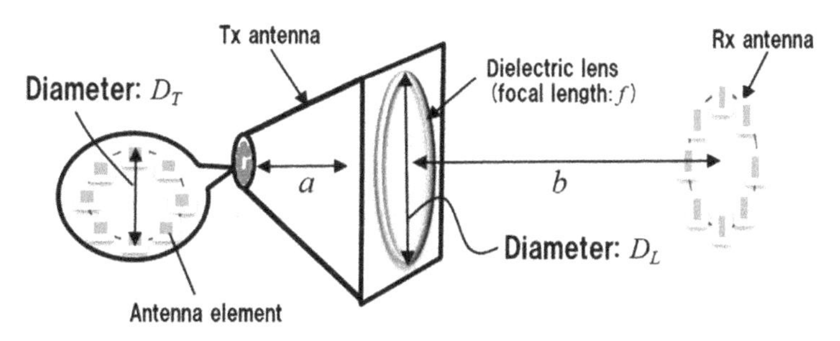

Figure 3. Configuration of the Tx and Rx antennas using dielectric lens.

By properly setting focal length f of the lens considering the distance between the lens and the transmitting antenna, the beam can be narrowed by the light converging effect. For example, Fukumoto et al. [4,18] showed that the beam spread can be reduced based on the imaging magnification at the reception point.

3.2. Experiments of OAM Multiplexing Using Dielectric Lens

To confirm the basic operation produced in using a dielectric lens, we performed a propagation experiment using the 28 GHz band in an anechoic chamber. Figure 4 shows the experimental setup. The measurement frequency of the network analyzer (NWA) was set from 2 to 3 GHz. Radio frequency (RF) chains up-converted 28–29 GHz inserted signals and fed them to the Tx antenna. At the receiver side, the received signals were down-converted to 2–3 GHz signals by the RF chains attached directly after the Rx antenna. We used a millimeter wave commercial lens whose directional gain and beam half-width were, respectively, 35 dBi and 3 cm (centimeters). The design requirements for dielectric lenses, including focal length, refractive index, diameter, directional gain, and beam half-width, remain open for analysis.

In the propagation measurements, we measured the channel coefficient between the Tx and Rx antennas using NWA while operating the positioner to move the Tx and Rx antennas to form the Tx and Rx UCAs elements' location sequentially. In other words, we emulated Tx and Rx UCAs using a single Tx and Rx antenna by sequentially measuring the channel coefficients of two points and combined the entire measurements to obtain channel characteristics between the emulated Tx and Rx antennas. The gain for both Tx and Rx antennas element was 27 dBi. We obtained $M \times N$ channel matrix H with measured channel coefficients consisting of all combinations between N-points (Tx side) and M-points (Rx side). Subsequently, we performed a matrix operation corresponding to OAM mode generation on the measured channel matrix and evaluated the phase and intensity distributions. The amplitude and phase characteristics of the OAM mode n were obtained by multiplying the column vector that generates OAM mode n by obtained channel matrix H.

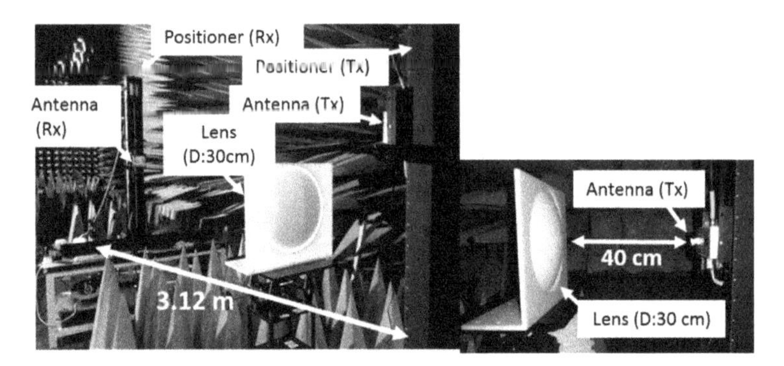

Figure 4. Experimental environment (OAM multiplexing using dielectric lens).

Table 1 shows experimental parameters. First, to confirm that an OAM mode was generated at the Rx side, we measured the phase distribution and the intensity distribution formed by the beam passing through the lens on the Rx side.

Table 1. Experimental parameters (Dielectric lens).

	Parameter	Value
Lens	Focal length: f	0.30 m
	Diameter: D_L	0.30 m
UCA	Number of antenna elements	12
	Diameter: D_T	0.04 m
	Distance between lens and UCA: a	0.40 m
Others	Frequency	28 GHz
	Distance between Tx and Rx: b	3.12 m

Figure 5 shows the measurement results and simulation results of the intensity distribution and phase distribution of OAM modes −2, −1, 0, +1, and +2. The measurements were conducted over a 60 cm × 60 cm grid and data were acquired at 3 cm intervals. The figure results confirmed that OAM beams whose phases rotate as much as their mode order were obtained in the experiments. The results also well matched the simulation results. Further, we consider that our experimental method will make it possible to emulate OAM beam generation.

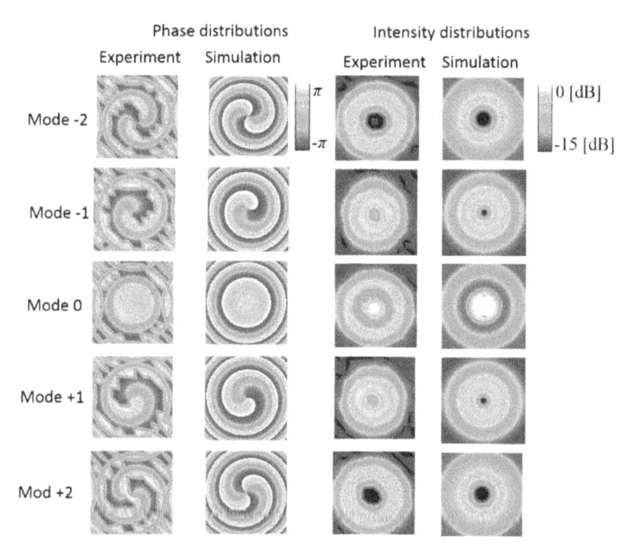

Figure 5. Measurement results and simulation results of phase and intensity distributions.

Next, we conducted experiments to ascertain the beam divergence reduction effect obtained with the lens. We took into account the distance from the center of the Rx side to the location of the strongest beam intensity since it is the quantitative index for evaluating beam divergence. We defined this metric as the beam diameter. The smaller is the beam diameter, the smaller is the beam divergence at the Rx side because the beam energy is more focused close to the center of the Rx antenna. Figure 6 shows comparisons between a UCA with a 4 cm diameter, a UCA with a 16 cm diameter, and a UCA with a 4 cm diameter plus a 16 cm lens.

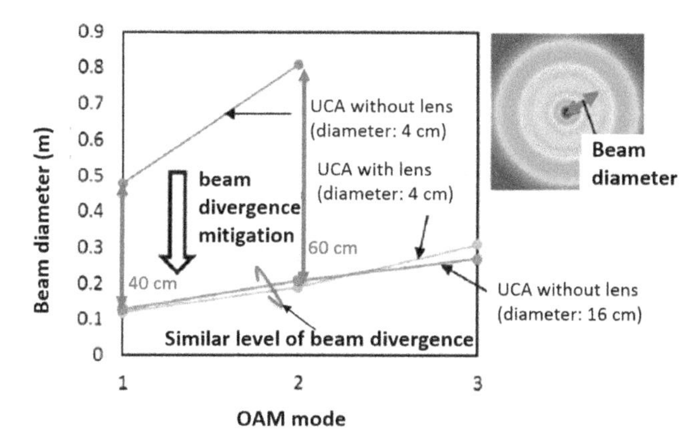

Figure 6. Experimental beam diameter results.

The beam diameters of the UCA with a lens in modes 1 and 2 were, respectively, 40 and 60 cm smaller than that of a UCA having a 4 cm diameter without a lens. We also found that the beam diameter of the UCA with a lens was in good agreement with the beam diameter of a UCA with a 16 cm diameter. These results enabled us to confirm that the beam divergence can be reduced by using a dielectric lens and that the transmission distance can be correspondingly increased. Experimental results also showed that aUCA with a 4 cm diameter and a lens produced an OAM beam similar to that produced by a UCA with a 16 cm diameter. These results suggested that using a dielectric lens is one of the options that will enable OAM wireless multiplexing systems to effectively address the beam divergence.

4. Experimental Demonstration Wireless OAM Multiplexing Technology using 28 GHz

4.1. *OAM Multiplexing Using Multiple UCAs*

The intensity distributions of OAM signals with different modes are given by the Bessel function of the first kind by their inherent nature, as shown in Figure 7. This yields a selective degradation when the Rx antenna is located in a low SNR region. To address this problem, we used multiple UCAs, which are designed to avoid reception in null regions. In this subsection, we first describe successful experimental results we obtained using five OAM modes over 28 GHz.

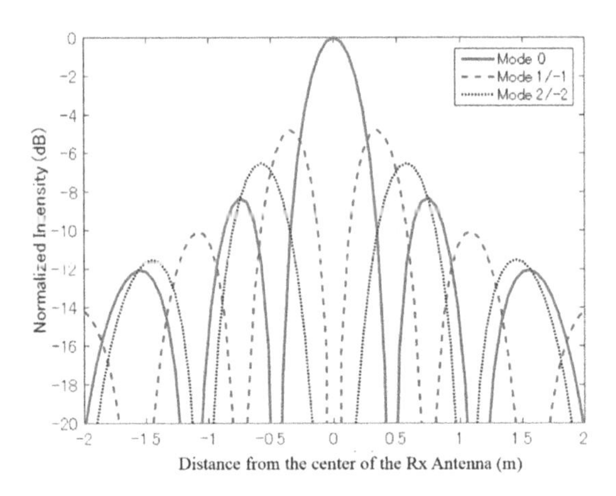

Figure 7. Bessel distributions of Rx signals.

We used multiple UCAs for both Tx and Rx antennas to rectify mode-selective Rx SNR degradation. Figure 8 shows our antenna design. It consists of four UCAs with different radii and a single antenna in the center. Each UCA consists of 16 antenna elements. The gain for each antenna element in each UCA was 11 dBi. The antenna in the center (hereafter UCA No. 0 for notation convenience) is used for the axis alignment and transmission of OAM mode 0. We used the following two methods to choose different UCAs for different OAM modes' transmission and chose Rx UCAs.

- Antenna Selection: Selecting a single Rx UCA that is not located at the null region of each OAM mode
- Receiver Diversity: Selecting multiple Rx UCAs to obtain Rx SNR enhancement by receiver diversity

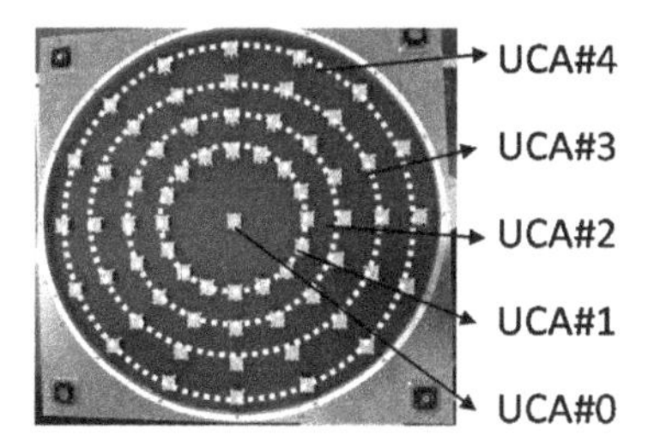

Figure 8. Implemented multiple uniform circular arrays (Four UCAs and a center antenna).

4.2. OAM-MIMO Multiplexing Using Multiple UCAs

This subsection presents wireless OAM-MIMO multiplexing that combines the OAM and MIMO concepts. The number of usable OAM modes is limited due to their mode-dependent beam divergence. In particular, higher OAM modes have a practical limitation due to their nature of large attenuation caused by beam divergence. To address this problem, we present OAM-MIMO multiplexing using multiple UCAs. Unlike OAM multiplexing, OAM-MIMO multiplexing exploits multiple sets of the same OAM modes with receiver equalizations. Consequently, the number of concurrently transmitted data streams can be increased without using higher OAM modes that have large beam divergence.

To enable OAM-MIMO multiplexing, we also used multiple UCAs such as OAM multiplexing. To achieve superposition-based simultaneous OAM beam generation and separation, we, respectively, implemented wideband analog 5×16 and 16×5 Butler matrices for Tx and Rx UCAs.

4.3. Experimental OAM Multiplexing Results Using Multiple UCAs

We conducted wireless OAM multiplexing experiments using five different OAM modes (−2, −1, 0, 1, and 2) over 28 GHz. Detailed descriptions regarding experimental parameters are given in Table 2.

Table 2. Experimental parameters (OAM multiplexing).

Parameter	Value	Parameter	Value
Center frequency	28.5 GHz	OAM modes	−2, −1, 0, 1, 2
Signal bandwidth	2 GHz	Number of streams	5
Number of UCAs	4	Signal carrier	Single carrier
Number of antenna elements in a UCA	16	Modulation	64 QAM [*1]
Number of antenna elements	65	Channel coding	LDPC (DVB-S2) 3/4
Diameter of UCA	24, 26, 48, 60 cm	Equalization	Frequency domain equalization
Tx/Rx distance	2.5 m	Block size	256

[*1] T quadrature amplitude modulation, [*2] low density parity check, [*3] digital video broadcasting satellite second generation.

We used the single carrier with frequency domain equalization (SC-FDE) to average channel characteristics over a wide signal bandwidth. The signal bandwidth was 2 GHz and 64-QAM modulation was used for all five OAM modes. The transmission rate per single stream was 9 Gbps. Experiments were conducted in a shielded room, as shown in Figure 9. The distance between Tx and Rx antennas was 2.5 m. The propagation loss can be calculated with Equation (2) with UCA sizes and distance between Tx and Rx. Tx signals were generated by arbitrary waveform generators and fed into a Tx antenna. The Rx antenna received signals and fed them into a digital oscilloscope that worked as an analog-to-digital convertor. Since the digital oscilloscope is equipped with four ports, reception was done sequentially using the four ports. This is not significant since the channel environments was static in the shield room. Offline digital signal processing was conducted using all the converted signals.

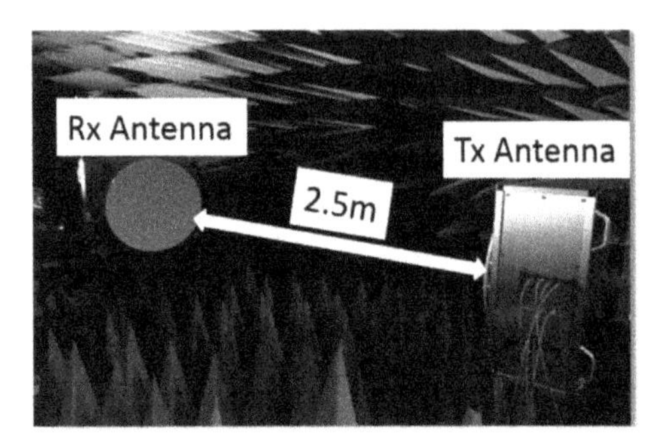

Figure 9. Experimental environment (OAM multiplexing).

Table 3 shows the experimental results we obtained for two methods. For both antenna selection and receiver diversity, we, respectively, used UCA No. 4, UCA No. 2, UCA No. 0, UCA No. 3, and UCA No. 1 for transmitting OAM modes −2, −1, 0, +1, and +2 signals. For the antenna selection method, we selected UCA No. 3, UCA No. 4, UCA No. 0, UCA No. 3, and UCA No. 1. By applying the LDPC channel coding (rate 3/4), we confirmed that error-free transmissions were obtained except for the OAM mode 2 case. For the receiver diversity method, we, respectively, selected UCA No. 4 and UCA No. 2 for OAM modes −1 and +2. We were able to confirm that error-free transmissions were received in all OAM modes with the same LDPC channel coding in this case. Using 2 GHz of signal bandwidth yielded a 45 Gbps transmission rate.

Table 3. Experimental results (OAM multiplexing).

		Mode −2	**Mode −1**	**Mode 0**	**Mode +1**	**Mode +2**
Antenna selection	Tx antenna	UCA No. 4	UCA No. 2	UCA No. 0	UCA No. 3	UCA No. 1
	Rx antenna	UCA No. 3	UCA No. 4	UCA No. 0	UCA No. 3	UCA No. 1
	BER (raw)	0.0114	0.0228	0.0201	0.0192	0.0428
	BER (coded)	0.0000	0.0000	0.0000	0.0000	0.0024
Receiver diversity	Tx antenna	UCA No. 4	UCA No. 2	UCA No. 0	UCA No. 3	UCA No. 1
	Rx antenna	UCA No. 3	UCA No. 2, No. 4	UCA No. 0	UCA No. 3	UCA No. 1, No. 2
	BER (raw)	0.0105	0.0015	0.0022	0.0278	0.0385
	BER (coded)	0.0000	0.0000	0.0000	0.0000	0.0000

4.4. Experimental OAM-MIMO Multiplexing Results Using Multiple UCAs

We conducted experiments on the OAM-MIMO multiplexing using our implemented antennas. Figure 10 and Table 4 show the experimental setup and parameters. Tx signals were generated by offline digital signal processing and fed into our implemented Tx antenna. Outputs of the Rx signals were fed into a digital oscilloscope that worked as an analog-to-digital convertor as in the OAM multiplexing experiments. In these experiments, the digital oscilloscope was equipped with four ports and reception was done sequentially using the ports. This also did not significantly affect the results since the channel environments were static in the shield room as in the OAM multiplexing experiments.

Figure 10. Experimental environment (OAM-MIMO multiplexing).

Table 4. Experimental parameters (OAM-MIMO multiplexing).

Parameter	Value	Parameter	Value
Center frequency	28.5 GHz	OAM modes	−2, −1, 0, 1, 2
Signal bandwidth	2 GHz	Number of streams	11
Number of UCAs	4	Signal carrier	Single carrier
Number of antenna elements in a UCA	16	Modulation	16 QAM/64 QAM
Number of antenna elements	65	Channel coding	LDPC (DVB-S2) 3/4, 5/6, 9/10
Diameter of UCA	24, 26, 48, 60 cm	Equalization	Freq. domain equalization
Tx/Rx distance	10 m	Block size	256

We used eleven streams for the OAM-MIMO multiplexing experiment. Two Tx UCAs (UCA No. 1 and No. 4), respectively, transmitted five OAM modes (0, ±1, ±2) and UCA No. 0 transmitted OAM mode 0. We used the received signals of all Rx UCAs for the equalization. Table 5 summarizes the modulations, channel coding rates and corresponding transmission rates used in the experiment. The propagation loss was calculated with Equation (2) with UCA sizes and distance between Tx and Rx. We confirmed that successful error-free transmissions were obtained in all streams with the usage of the forward error correction. Total transmission rate was 100 Gbps at a 10 m transmission distance. The results indicated a new milestone was reached in terms of point-to-point wireless transmission. In addition, we recently extended our work by extending the baseband signal processing to successfully achieve 120 Gbit/s using a total of 11 OAM modes [19]. This is the state-of-the-art results that have been published in the literature.

Table 5. Experimental results (OAM-MIMO multiplexing).

		Mode −2	Mode −1	Mode 0	Mode 1	Mode 2
Tx UCA No. 0	Modulation (QAM)			64		
	Channel coding rate			3/4		
	Trans. rate (Gbps)			9		
Tx UCA No. 1	Modulation (QAM)	64	64	16	64	64
	Channel coding rate	9/10	3/4	9/10	9/10	5/6
	Trans. rate (Gbps)	10.8	9	7.2	10.8	10
Tx UCA No. 4	Modulation (QAM)	64	64	16	64	64
	Channel coding rate	3/4	3/4	9/10	3/4	3/4
	Trans. rate (Gbps)	9	9	7.2	9	9

5. Conclusions

In this paper, we describe how we investigated wireless orbital angular momentum (OAM) multiplexing performance from theory to experimental perspectives. First, we conducted basic theoretical studies on wireless OAM multiplexing performance with respect to various aspects including modulation, demodulation, multiplexing, and demultiplexing. Then, we expanded our interest to experimental evaluation including a dielectric lens and end-to-end wireless transmission at 28 GHz frequency bands. We confirmed that using a dielectric lens can effectively reduce the beam divergence effect and correspondingly increase the transmission distance. In end-to-end experiments, we achieved 45 Gbps throughput using five OAM modes. In addition, we experimentally demonstrated the effectiveness of our proposed OAM-MIMO multiplexing method using a total of 11 OAM modes. In the experiments, we reached a new milestone in point-to-point transmission rates by achieving 100 Gbps at a 10 m transmission distance.

Author Contributions: D.L. and H.S. contributed to conceptualization and methodology, H.S. and H.F. conducted validation, D.L., H.S., H.F. and Y.Y. conducted data curation, D.L. contributed to writing original draft preparation and editing, T.S. supervised overall work.

Funding: This research received no external funding.

Conflicts of Interest: The authors declare no conflict of interest.

References

1. Yan, Y.; Xie, G.; Lavery, M.P.J.; Huang, H.; Ahmed, N.; Bao, C.; Ren, Y.; Cao, Y.; Li, L.; Zhao, Z.; et al. High-capacity millimeter-wave communications with orbital angular momentum multiplexing. *Nat. Commun.* **2014**, *5*, 4876. [CrossRef] [PubMed]
2. Lee, D.; Sasaki, H.; Fukumoto, H.; Hiraga, K.; Nakagawa, T. Orbital Angular Momentum (OAM) Multiplexing: An Enabler of a New Era of wireless Communications. *IEICE Trans. Commun.* **2017**, *E100-B*, 1044–1063. [CrossRef]
3. Yan, Y.; Li, L.; Zhao, Z.; Xie, G.; Wang, Z.; Ren, Y.; Ahmed, N.; Sajuyigbe, S.; Talwar, S.; Tur, M.; et al. 32-Gbit/s 60-GHz millimeter-wave wireless communication using orbital angular momentum and polarization multiplexing. In Proceedings of the 2016 IEEE International Conference on Communications (ICC), Kuala Lumpur, Malaysia, 22–27 May 2016; pp. 1–6.
4. Lee, D.; Sakdejayont, T.; Sasaki, H.; Fukumoto, H.; Nakagawa, T. Performance evaluation of wireless communications using orbital angular momentum multiplexing. In Proceedings of the 2016 International Symposium on Antennas and Propagation (ISAP), Okinawa, Japan, 24–28 October 2016.
5. Fukumoto, H.; Lee, D.; Sasaki, H.; Kaho, T.; Shiba, H. An experimental study on beam focusing effect using dielectric lens for OAM multiplexing. *IEICE Tech. Rep.* **2018**, *117*, 53–57.
6. Lee, D.; Sasaki, H.; Fukumoto, H.; Yagi, Y.; Kaho, T.; Shiba, H.; Shimizu, T. Demonstration of an orbital angular momentum (OAM) multiplexing at 28 GHz. *IEICE General Conf.* **2018**, B-5-90.
7. Lee, D.; Sasaki, H.; Fukumoto, H.; Yagi, Y.; Kaho, T.; Shiba, H.; Shimizu, T. An experimental demonstration of 28 GHz band wireless OAM-MIMO (orbital angular momentum multi-input multi-output) multiplexing. In Proceedings of the IEEE 87th Vehicular Technology Conference (VTC Spring), Porto, Portugal, 3–6 June 2018.
8. Mari, E.; Spinello, F.; Oldoni, M.; Ravanelli, R.A.; Romanato, F.; Giuseppe, F.; Parisi, G. Near-field experimental verification of separation of OAM channels. *IEEE Antennas Wirel. Propag. Lett.* **2015**, *14*, 556–558. [CrossRef]
9. Willner, A.E. Communication with a twist. *IEEE Spectrum* **2016**, *53*, 34–39. [CrossRef]
10. Mahmouli, F.E.; Walker, S.D. 4Gbps uncompressed video transmission over a 60-GHz orbital angular momentum wireless channel. *IEEE Wirel. Commun. Lett.* **2013**, *2*, 223–226. [CrossRef]
11. Cheng, L.; Hong, W.; Hao, Z. Generation of electromagnetic waves with arbitrary orbital angular momentum modes. *Scientific Rep.* **2014**, *4*, 4814. [CrossRef] [PubMed]
12. Jin, J.; Luo, J.; Zhang, X.; Gao, H.; Li, X.; Pu, M.; Gao, P.; Zhao, Z.; Luo, X. Generation and detection of orbital angular momentum via metasurface. *Scientific Rep.* **2016**, *6*, 24286. [CrossRef] [PubMed]
13. Deng, C.; Chen, W.; Zhang, Z.; Li, Y.; Feng, Z. Generation of OAM Radio Waves Using Circular Vivaldi Antenna Array. *Int. J. Antennas Propag.* **2013**, *2013*, 1–7. [CrossRef]

14. Haskou, A.; Mary, P.; Hélard, M. Error probability on the orbital angular momentum detection. In Proceedings of the 2014 IEEE 25th Annual International Symposium on Personal, Indoor, and Mobile Radio Communication (PIMRC), Washington, DC, USA, 2–5 September 2014; pp. 302–307.
15. Opare, K.A.; Kuang, Y.; Kponyo, J.J.; Nwizege, K.S.; Enzhan, Z. The degree of freedom in wireless line-of-sight OAM multiplexing system using a circular array of receiving antenna. In Proceedings of the 2015 Fifth International Conference on Advanced Computing & Communication Technologies, Rohtak, Haryana, India, 21–22 February 2015.
16. Opare, K.A.; Kuang, Y.; Kponyo, J.J.; Nwizege, K.S.; Enzhan, Z. Mode combination in an ideal wireless OAM-MIMO multiplexing system. *IEEE Wirel. Commun. Lett.* **2015**, *4*, 449–452. [CrossRef]
17. Tian, H.; Liu, Z.; Xi, W.; Nie, G.; Liu, L.; Jiang, H. Beam axis detection and alignment for uniform circular array-based orbital angular momentum wireless communication. *IET Commun.* **2016**, *10*, 44–49. [CrossRef]
18. Fukumoto, H.; Lee, D.; Sasaki, T.; Nakagawa, T. Beam divergence reduction using dielectric lens for orbital angular momentum based wireless communications. In Proceedings of the 2016 International Symposium on Antennas and Propagation (ISAP), Okinawa, Japan, 24–28 October 2016.
19. Sasaki, H.; Lee, D.; Fukumoto, H.; Yagi, Y.; Kaho, T.; Shiba, H.; Shimizu, T. Experiment on Over-100-Gbps Wireless Transmission with OAM-MIMO Multiplexing System in 28-GHz Band. In Proceedings of the 2018 IEEE Global Communications Conference (GLOBECOM), Abu Dhabi, United Arab Emirates, 9–13 December 2018.

MDPI
St. Alban-Anlage 66
4052 Basel
Switzerland
Tel. +41 61 683 77 34
Fax +41 61 302 89 18
www.mdpi.com

Applied Sciences Editorial Office
E-mail: applsci@mdpi.com
www.mdpi.com/journal/applsci

CPSIA information can be obtained
at www.ICGtesting.com
Printed in the USA
LVHW051005280120
645028LV00010B/223

9 783039 212231